THEMES AND DIMENSIONS
— OF THE —
NATIONAL
CURRICULUM

Kogan Page Books for Teachers series
Series Editor: Tom Marjoram

THEMES AND DIMENSIONS

— OF THE —

NATIONAL CURRICULUM

Implications for Policy and Practice

EDITED BY

GEOFFREY HALL

KOGAN
PAGE

First published in 1992

Apart from any fair dealing for the purposes of research or private study, or
criticism or review, as permitted under the Copyright, Designs and Patents
Act, 1988, this publication may only be reproduced, stored or transmitted, in
any form or by any means, with the prior permission in writing of the
publishers, or in the case of reprographic reproduction in accordance with
the terms of licences issued by the Copyright Licensing Agency. Enquiries
concerning reproduction outside those terms should be sent to the
publishers at the undermentioned address:

Kogan Page Limited
120 Pentonville Road
London N1 9JN

© Geoffrey Hall and named contributors, 1992

British Library Cataloguing in Publication Data

A CIP record for this book is available from the British Library.

ISBN 0 7494 0555 4

Typeset by DP Photosetting, Aylesbury, Bucks
Printed and bound in Great Britain by
Biddles Ltd, Guildford and Kings Lynn

Contents

List of Contributors

Geoffrey Hall is Associate Director of the Evaluation and Assessment Unit in the Department of Education at Liverpool University. He is the author of *Records of Achievement: Issues and Practice* (Kogan Page 1989), and was previously Curriculum Development Officer with the Metropolitan Borough of Knowsley.

Alan Blyth was Professor of Education at Liverpool University and is still an Honorary Senior Fellow there. He taught in schools and colleges and in the Universities of Keele and Manchester. His main interests have been in primary education, especially the role of the humanities in the primary curriculum in which he directed a major curriculum development project in the 1970s. He has been involved in many recent developments in primary industry education, including the National Curriculum Council's Task Group on Economic and Industrial Understanding.

Jim Burke is currently TVEI(E) Coordinator for the Metropolitan Borough of Knowsley. Formerly Head of the Careers Education and Guidance Department in a large comprehensive school, he has spent a period on secondment to the Borough's Careers Service.

Paul Davies is at present Advisory Teacher for Humanities in the Metropolitan Borough of Knowsley. Previously he was the Borough's Co-ordinator for Multicultural Education and was Head of History in a Merseyside comprehensive School.

Ray Derricott is a Reader and Director of Continuing Education at the University of Liverpool. He has just completed four years' involvement in the ESRC 16–19 Initiative, a national project investigating the lifestyles, attitudes and beliefs of a sample of young people in Kirkcaldy, Liverpool, Sheffield and Swindon. As Director of the Liverpool

Evaluation and Assessment Unit, he is responsible for the evaluation of the Enterprise in Higher Education Initiative within the University. He also heads the national evaluation of the use of a Personal Competences Model which is being piloted in schools, colleges of further education, higher education institutions and in work situations in England, Scotland and Wales.

Cliff Jones works in the Assessment Team of Liverpool Education Authority. His work is centred around assessment, accreditation and records of achievement. Currently his main interest is in establishing a system involving employers, trade unions, parents, higher education institutions, students and others in endorsing and accrediting the quality of learning.

Pat King has been an Advisory Teacher for PSE and Health Education with the Metropolitan Borough of Knowsley since 1986, and was previously Head of Biology in a large secondary school.

Ann Mabbott has taught in secondary education since 1964 and is currently Deputy Headteacher of a comprehensive school in the Sefton Local Education Authority. Between 1989 and 1991, she was seconded by the LEA to draft a policy and strategy document in relation to Equal Opportunities and the National Curriculum.

Bill Marsden is Reader in Education at the University of Liverpool and was until recently Dean of its Faculty of Education and Extension Studies. His publications include, *Unequal Educational Provision in England and Wales: the Nineteenth-century Roots* and *Educating the Respectable: a Study of Fleet Road Board School, Hampstead, 1879–1903*.

Janet Strivens is a lecturer in education at Liverpool University. She teaches at all levels of the department's work with a particular interest in learning strategies and the development of social, moral and political understanding. She is currently involved in the development of enterprise skills in higher education and initial teacher training.

David Thomas is Senior Lecturer in Education and Director of the Professional Development and In-Service Unit at Liverpool University. He is the author of *An introduction to the literature of special education. The social psychology of childhood disability* and *The experience of handicap*.

David Wheatley Following 14 years' experience as a lecturer in biological sciences at the St Helens Community College, David was appointed Environmental Education Development Officer for the Mersey Basin Campaign Voluntary Sector Network in June 1989. Based at the University of Liverpool Department of Education, he has directed numerous courses and seminars on environmental education and has had practical experience in primary and secondary schools throughout the North-West of England.

Acknowledgements

I would like to thank my co-authors for their valued contributions, co-operation and support; also Linda Gill for her efficiency and patience in co-ordinating the production of the manuscripts and index.

Geoffrey Hall
Editor

Introduction

Does the Curriculum Match the Reality?

Geoffrey Hall

A brief glance at the daily papers and television news informs us of the importance that is attached to issues which have become prominent national and international concerns. These include, among many others, problems related to health, the environment, equal opportunities, international understanding and, almost certainly, the economy and industrial growth. If these matters are of such importance then it would seem logical that the curriculum should reflect this priority.

It could be argued that Section 1 of the Education Reform Act 1988 (ERA) was designed to provide the legislative framework for this to happen. The ERA

> places a statutory responsibility upon schools to provide a broad and balanced curriculum which promotes the spiritual, moral, cultural, mental and physical development of pupils . . . and prepares pupils for the opportunities, responsibilities and experiences of adult life (NCC, 1990a)

The whole curriculum is intended to serve these aims and is defined as including, among other elements, the National Curriculum augmented by religious education (RE), and 'an accepted range of cross-curricular elements' (NCC, 1990a).

The major component is obviously the National Curriculum with RE and this is known as 'the basic curriculum'. However, it is accepted that this is inadequate in itself to deliver the knowledge, skills and understanding required to cover the wide curriculum areas identified above. With regard to the cross-curricular elements, they are seen as 'ingredients which tie together the broad education of the individual and augment what comes from the basic curriculum' (NCC, 1990a). The National Curriculum Council (NCC) considers the cross-curricular elements to have three aspects: themes, dimensions and skills. The

themes include economic and industrial understanding, careers education and guidance, health education, education for citizenship and
environmental education, while the dimensions are concerned with a
commitment to providing equal opportunities for all pupils and
recognition that preparation for life in a multi-cultural society should
permeate every aspect of the curriculum. The skills comprise communication, numeracy, study, problem-solving, personal and social and
information technology.

The important difference between the elements and the National
Curriculum plus RE is that the former is non-statutory while the latter
is statutory. This has led to the National Curriculum being the
Government's priority for implementation. Moreover, this status has
been reinforced by the introduction of an elaborate system of assessment. On the other hand, the cross-curricular themes and dimensions
are only statutorily assessed where they permeate the National
Curriculum subjects. Furthermore, it is difficult to assess an individual
theme because rarely, if ever, would it be taught as a separate entity,
such is the dominance of the National Curriculum. Yet as the opening
of this chapter indicated, there are topics of national and international
importance which, on reflection, relate more obviously to the themes
and dimensions than to the National Curriculum subjects. The
situation is, then, that while they may reflect many of the everyday
experiences of the general public, it is difficult to identify their place in
the National Curriculum unless a mapping exercise is carried out across
the subjects. This monitoring process must also take place in the
remainder of the whole curriculum with the additional administration
burden this entails. To sum up, the themes and dimensions cover
important areas of knowledge and understanding, but they are non-
statutory and may not appear as a coherent whole, because individually
they are likely to be spread across the whole curriculum.

With regard to this latter point the same can be said for the
assessment arrangements, only the position is made even more difficult
because of the differences between statutory and non-statutory
practices.

The cumulative effect of this curriculum organisation has been to
lower the status and therefore the priority of the themes and
dimensions in the eyes of the teaching profession as a whole. The
teachers have been so hard pressed implementing the National
Curriculum and other aspects of the ERA that it is not surprising that
wider issues have received only limited attention. The purpose of this
book, therefore, is to re-assert the importance of access to and
implementation of a curriculum which more usefully reflects the world
young people are experiencing and are soon to enter as adults. In doing

so it points up the limitations of the National Curriculum in terms of achieving the aims of the ERA and argues that in the end it will be a challenge to the values held by teachers as to whether this broad approach is implemented. They will require strong constitutional support in a climate which promotes conformity to narrowly conceived goals.

In order that some coherence can be given to the curriculum approaches envisaged, the contributors have taken as their starting points the themes and dimensions identified by the NCC and referred to above. To commence, Alan Blyth uses a parable to illustrate how the National Curriculum appears to have been derived. He traces the emergence of the hierarchical subject basis of the National Curriculum and then the cross-curricular elements. He goes on to analyse their principal characteristics, and to argue for a whole curriculum based upon the three modes of understanding and experience, action and purpose with the themes and dimensions being central rather than peripheral.

In Chapter 2, David Wheatley makes the case for environmental education by establishing the need and illustrating how it might be met by producing case studies of primary and secondary schools, and an example of partnership with local industry. He also points to the important role of voluntary organisations in this work. He concludes by suggesting how the environmental cause might be promoted through careers and health education and citizenship.

Pat King, in Chapter 3, analyses the rationale underpinning health education, and explores, in particular, the NCC's Curriculum Guidance 5 on this subject. Writing on the basis of considerable experience in primary and secondary schools, she evaluates various approaches to implementing health education and offers examples of good practice.

Chapter 4 focuses on careers education and guidance across the curriculum. While this theme may not represent headline news, in the same way as for example international understanding with respect to developments on the European scene, its importance cannot be over-estimated. Once a 'cinderella' area in the secondary curriculum, it is now nationally accepted as being vital 'in promoting self-awareness and . . . is a pre-requisite to making well-informed educational and training choices' (NCC, 1990b). Recently, careers guidance has come to the fore with student action-planning being seen as a vital component of the work placement schemes of Training and Enterprise Councils (TECs). Jim Burke's essentially practical approach commences with a definition of the subject, followed by a consideration of its rationale and content, and review of the literature. He concludes with step-by-step case studies of the development of careers education policies in a primary and a secondary school.

Chapters 5 and 6 are devoted to economic and industrial understanding (EIU). First, Alan Blyth examines approaches to this theme for the 5–13 age group, focusing on the idea of the 'Mini-market'. After challenging the notion of EIU as a unitary concept, the meanings of 'economic' and 'industrial' are teased out. This is followed by a discussion of the arguments about the Mini-market idea, leading to a consideration of curriculum planning issues for National Curriculum Key Stages 1–3.

Second, Ray Derricott, in Chapter 6, examines teaching EIU at National Curriculum Key Stage 4 and beyond. This focus was chosen because there has been a variety of initiatives in this sector and the question of curriculum continuity beyond the statutory school leaving age of 16 is an important issue. These matters are then discussed in relation to the literature and organisations involved. However, it is important to examine how this debate is received on 'the ground', so the author details and analyses interviews conducted with two heads using evidence from a Business/Education Partnership Project conducted in north-west England. The issues which arise are then considered in terms of where future developments may lie, informed by an outline of the TVEI 16–18 Enrichment Project Pilot currently based at the University of Liverpool.

International understanding (IU), it could be argued, needs little justification to have an important place in the whole curriculum. Recent events in Europe, the Middle East and the developing world strongly support this view. Yet to identify its place requires the kind of mapping exercise referred to earlier. The most prominent reference is to be found under the heading of 'roles and relationships in a pluralist society' in the NCC's Curriculum Guidance 8, Education for Citizenship. Bill Marsden, in his contribution (Chapter 7), argues that the curriculum has been 'nationalised', and that the international dimension has been marginalised. He suppports this assertion by an analysis of history, modern languages, geography and the cross-curricular themes and dimensions in this respect and paints a pessimistic scenario. However, he goes on to assert that geography provides the best opportunities for developing international understanding. He offers four examples of good practice together with titles of useful reference material to reassure teachers that, despite its limitations, the National Curriculum gives opportunities to develop international understanding.

On the face of it, the concept of citizenship should bring together many of the aspects of the cross-curricular elements, because much of what has been included would be considered to be within the definition of 'good citizenship'. However, the NCC's Curriculum Guidance 8 widens this concept to include 'eight essential components' which go beyond the scope of the other themes and dimensions. Janet Strivens and Cliff Jones discuss citizenship in the context of the National Curriculum and the wider political scene. They go on to suggest how it might be delivered and the requirements for good citizenship education in terms of teacher education, the schools and society.

The subject of the final chapter, on equal opportunities (EO) within the National Curriculum, underpins, everything that has gone before. The authors commence with an analysis of the National Curriculum literature with respect to EO and the curriculum models available to teachers. The place of EO is then explored in the context of race, gender and special educational needs (SEN) with respect to the paradox of Sections 1 and 2 of the ERA, the response of the schools, assessment arrangements and the limitations of the ERA in addressing SEN. The National Curriculum core and foundation subjects and RE are analysed in terms of EO. Finally, practical advice is given with regard to implementation in the form of a case study of school-focused in-service provision.

In conclusion, this book has aimed at promoting a broad view of curriculum entitlement. This position is strongly supported by HMI in an overview of the curricular requirements of the ERA. They argue:

> In planning and implementing the curriculum, teachers along with governing bodies should not lose sight of particular aims and objectives which fall outside the scope of the Core and other Foundation Subjects. They will need to consider not just other subjects but also a range of themes, dimensions and skills that are in many more than one area of the curriculum (HMI, 1991).

In contrast, the Secretary of State for Education and Science has welcomed the view that subject-based teaching should be more widely introduced in the 9 to 11 age range (Hart, 1991), and the NCC in a recent survey, noted that cross-curricular issues were losing out to the National Curriculum in secondary schools. This situation has been compounded by the Secretary of State's distrust of 'child-centred' learning and his decision to set up a committee of enquiry into primary teaching methods and class organisation (TES, 1991).

Given this climate, together with the many examples in this book of the pressures upon teachers when faced with the difficulties of establishing cross-curricular themes and dimensions in the whole curriculum, it would not be surprising it they found the task daunting. It is their commitment to these values which is of critical importance. To succeed they will need considerable support from their institutions, local authorities, and, most important, a sympathetic response from Her Majesty's Government.

REFERENCES

Hart, D (1991) 'An Agenda for Primary Education for the 1990s', speech by David Hart OBE, General Secretary, National Association of Head Teachers, 3 November.

HMI (1991) *The Implementation of the Curricular Requirements of the ERA; An Overview by H.M. Inspectorate on the First Year, 1989/90*, London: HMSO.

NCC (1990a) *Curriculum Guidance 3: The Whole Curriculum*, York: NCC.

NCC (1990b) *Curriculum Guidance 6: Careers Education and Guidance*, York: NCC.

The Times Educational Supplement, 6 December 1991.

Chapter 1

Themes and Dimensions: Icing or Spicing?

Alan Blyth

PARABLE

There was once a Queen who saw that everyone else had a cake and she had not.

So she sent for her Master Baker to ask what ingredients a really good cake ought to have. He in his turn enquired of one or two people, and then reported to the Queen that her cake ought to consist of the best flour, the finest sugar, top-quality eggs, some prime butter and some other goodies such as currants and sultanas and spices. So she ordered the Baker to bring together the people who made the best flour, the finest sugar, and the really up-market versions of all the other ingredients. They all brought the very best they had and delivered it to the Baker. Then he fired the oven and put in the ingredients one by one, beginning with the most important ones, and so the cake was baked. Not half-baked, for that would never do. If anything, it was baked too well.

But while it was being baked, some other people came to the Baker and said that they too had brought ingredients that ought to have been added to the cake. And while they were yet speaking, still other people ventured to say that the cake mixture should have been stirred very well before the baking began. But it was too late, and anyhow the Baker said that he knew what a cake should be made of.

All the same, when it came out of the oven, you could see that one part of it was like bread, another part like burnt sugar, and there were soft places where the eggs and the butter had gone in. Even the currants and sultanas were still all together, and you could smell the spices in some parts but not in others. Admittedly, along the edges between where the different ingredients had been, there was some sort of mixture, but that was all. However, the Baker told the Queen what a good cake it was, though he noticed how many people were shaking their heads, and that made him think.

Then he had one of the bright ideas for which he told people he was famous. If there could be an icing on the cake, that would surely put things right for everybody. The people who said that the mixture ought to have been stirred would be happy, because now nobody would notice. The people who wanted the other ingredients put in could have them in the icing and, since they had brought them already, there would be no need to buy anything else. So he had the icing put on the cake, and everybody was happy ever after, except for a few specialists in home economics, and who cares what they think, anyhow?

I do, for one. That is not the way to make a cake, and it is not the way to make a curriculum. But it is the way in which our National Curriculum was made, and we have to make do with it.

PRELUDE

At least, that is how our National Curriculum appears to have been made. We shall not know for certain until the archives are opened. The story ran, it seems, something like this.

When, during the 1980s, it became obvious that there was a tide running in favour of some sort of National Curriculum, various attempts were made to decide on what basis it should be constructed. HMI were among those who gave advice, suggesting that it might well be centred on *areas of learning and experience* (HMI, 1985). Others advanced the claims of particular subjects, or of groups of subjects. In the end there was a surge of agreement that, in spite of HMI, any curriculum, primary or secondary, ought to consist of 'subjects'. Indeed, the idea was assiduously fostered that nobody with any common sense could doubt that it should consist of subjects, and that anyone who thought otherwise was suspect of belonging to something called the 'Educational Establishment', identified as a prime cause of national decline. So subjects it should be, and the Education Reform Act, 1988, enshrined this principle in statute.

It soon became apparent that some subjects were weightier than others. In the Act, and in subsequent administrative elaborations, a kind of caste system of subjects emerged. At the top stood the 'old basics', known affectionately as the 'three Rs' (and one W, for Welsh in parts of Wales). Almost, but not quite, equal in esteem came the 'new basics': science, with the accolade of a core subject, and technology – the universal acceptance of technology was a remarkable phenomenon that will require further attention later; it just failed to reach core status. On the third tier stood history and geography, apparently in a love–hate relationship, and modern (European) languages, along with Welsh in the rest of Wales. Art, music and physical education were grouped together, one step down. Religious education was allowed a unique position, being in the Basic Curriculum but not in the National Curriculum (not inappropriately, if religion is something basic but not national). That still left others outside the caste system altogether, a strange mish-mash of academic refugees including classical languages, drama, economics, sociology, home economics and whatever. Although not in the National or even the Basic Curriculum, they were not actually excluded from the wider whole curriculum, but in practice the Basic Curriculum and even the National Curriculum bid fair to occupy a high percentage of the whole curriculum: some estimates put it at about 110 per cent.

This caste system did appeal to some people, especially those in the highest castes, but of course it offended other people, especially those in the lowest castes, including the academic refugees. It also offended those who – rightly – would have preferred to start from the whole curriculum and to involve subjects as they became relevant, rather than the other way round. So there were two groups of dissidents, and in practice they formed quite an effective coalition of opposition. They were organised in different pressure groups, some of them established and distinguished, others new and assertive. When (also under the Education Reform Act) the National Curriculum Council was set up, it was already clear that some attention would have to be paid to what these groups were saying. Following ideas adumbrated by HMI and others, the notion took root that the concerns of objectors might be accommodated under *cross-curricular elements* which would blend with, and be nourished by, subjects, and would thus have the great additional merit of making no calls upon additional time or resources. Although the earlier publications *National Curriculum: From Policy to Practice* (DES, 1989) and *A Framework for the Primary Curriculum* (NCC, 1989a) observed uncommitted banality on this issue, pending further developments, the first of these did indicate (para. 4.7) that 'The Secretary of State has asked NCC to give urgent attention . . .' to cross-curricular considerations. So urgent was this attention that, in the event, it is as though a few clever people were locked in a room one afternoon and told to come out with an answer. That is when the cake was iced.

The terminology was not entirely settled at once but, through the issue by the NCC of *Circular 6* (NCC, 1989b), followed by the fuller *Curriculum Guidance 3* (NCC 1990), the current pattern became established:

- *Themes* – the elements intended to have some positive content:
  economic and industrial understanding
  health education
  careers education and guidance
  environmental education
  education for citizenship.
- *Dimensions* – more closely related to personal development:
  multicultural considerations
  gender issues
  special educational needs.
- *Skills* – originally 'Competences':
  numeracy (including graphicacy)
  problem-solving skills
  information-handling skills
  study skills.

The choice of these three categories of cross-curricular elements was necessarily arbitrary. Themes can certainly claim some curricular ancestry, and skills are generally recognised as being important and relevant, though some of these skills are more cross-curricular than others and some look very like the attainment targets in the old basics. Meanwhile, information technology, originally included as a cross-curricular skill, was eventually promoted to the status of an attainment target, indeed a whole profile component, in technology: part of the cake (and a good one) instead of part of the icing. However, the third term, dimensions, seems oddly chosen. There is no question that the three instances specified should be essential elements in any curriculum, but it seems artificial to call them *dimensions*. How long, or broad, or tall are they supposed to be? Was the term chosen because there are three of them? Could they not have been called facets, or threads, or, to use a term fashionable just before the National Curriculum was born, *permeations*? However, the terminology is now set, at least for the time being.

Themes and dimensions are the subject of this book, and will be considered in subsequent chapters. However, the actual selection of themes and dimensions, and the way in which they are perceived, calls for a little more analysis.

THEMES

Before these five come to be regarded as immutable, each with its own *Curriculum Guidance* to ennoble it, their very tentative origin should be remembered. As *Curriculum Guidance 3* put it:

The National Curriculum Council has identified five themes which, although by no means a conclusive list, seem to most people to be preeminent. It is reasonable to assume at this stage that they are essential parts of the whole curriculum (NCC, 1990, p. 4).

Each phrase in this quotation would in fact merit an exegesis of its own, to consider the many underlying assumptions and implications. A little later, the same document declares: 'It must remain open to schools to decide how these themes are encompassed within the whole curriculum' (p. 6).

The first of these quotations surely defies anyone to challenge such tolerant wisdom from so eminent a source, while the second deftly transfers to schools the task of actually weaving the cross-curricular themes into their programmes, presumably 'within existing resources'. (For the resulting problems see Hargreaves, 1991.)

Nevertheless the five themes managed to satisfy a lot of people: economists who saw a chance for their subject; social scientists who found a niche in at least three of them; biologists who, perhaps outweighed by the physical scientists, came into their own in two; and a whole clutch of interest groups who saw their way to influencing the curriculum in their direction by means of one or more of the themes.

As *Curriculum Guidance 3* avers, there is overlap between these somewhat disparate themes. Careers education and guidance does not part company meaningfully from economic and industrial understanding until Key Stage 3 or even 4. Environmental education actually overlaps with one of the attainment targets in geography. Education for citizenship and health education are both components of personal/social education, which is related to the dimensions as well as the themes. Both of these have also an international or global aspect, which might indeed be developed into a sixth theme, international understanding, as is suggested later in this book. None of these considerations need cause difficulties, provided that the relations between the themes and the dimensions and the various subjects is clearly borne in mind. This is in fact spelled out not only in the *Curriculum Guidance* series but also in the Final Orders for all the core and other foundation subjects themselves. There is no lack of prompting to forge links across the curriculum. All of these documents elaborate that agenda in increasing detail. But it is still left to teachers, assisted where possible by the shrinking advisory service and hard-pressed institutions of higher education, to do the forging of those links in their particular situations. This is one piece of autonomy that the Department of Education and Science is only too grateful to leave to teachers.

DIMENSIONS

As has already been indicated, this is the more problematic category. Originally personal/social education was to have been included, but eventually it became something so all-embracing that it was deemed to include all of them and more besides. So the list became confined to three kinds of potential bias and discrimination, each of which has a dual significance, being concerned partly with the content of the curriculum (is this biased, or does it appear to be against, or for, any social or cultural group?) but also with the 'delivery' of the curriculum (is equal, or at any rate fair, access afforded to the curriculum itself?).

The problems arise when the remaining three dimensions are considered, and later chapters will bring these problems to the fore. Some of the related issues have become familiar in the public domain. The most controversial, probably, is whether the multicultural dimension should, or must, be anti-racist, and what are the policy implications of any answer to that question. Almost as sharp are the questions relating to gender, which had originally been excluded from the list, ostensibly on grounds of its imprecision (Miles and Middleton, 1990, p. 200). In practice, the terms 'gender' and 'equal opportunities' have been used almost interchangeably, but that verbal finesse does not mean that the two concepts are identical or that they both presage the same policies. With special educational needs, the situation is a little different. There is a long-standing recognition that the curriculum should be adapted for children unable or unwilling to take part in the usual school programmes and, uniquely among dimensions, special educational needs was allowed a *Curriculum Guidance* document of its own (NCC, 1989c). Yet there is plenty of room for disagreement about which children fall into this category, how far the National Curriculum should be 'disapplied' in their case (as though it could be switched off like a power supply), and what should happen after disapplication and before what must be termed, with resonances of the job market, 're-application'. All these matters are vitally important, even if they are rather dubiously termed 'dimensions'.

A MODAL CURRICULUM

It would have been so much better if the process of making the National Curriculum had started from the whole curriculum, primary and secondary, instead of from subjects. Or, if a start from the whole curriculum seems too nebulous, at least there could have been a start from the cross-curricular elements themselves, as in conflict-ridden

Northern Ireland where the consideration of the various subjects *followed* the designation of cross-curricular themes such as EMU (Education for Mutual Understanding).

Perhaps we could usefully think of a curriculum based on three equally important kinds of *modes*: modes of understanding and experience, modes of action and modes of purpose. In such a curriculum, the present themes and dimensions would be much more adequately represented.

Modes of understanding and experience

The first of these modes would indeed be concerned with subjects, but not in the same way as they are found in a core-plus-foundations structure. For subjects do matter. A curriculum without subjects, even – or especially – if it is made by children themselves, would be like a cake without ingredients. But that is the point: they are ingredients that need to be blended in, not building blocks that have to be erected alongside or on top of each other. Another way of describing subjects is that they are *resources* or *perspectives*, each based on one or more disciplines that represent distinctive areas of human achievement. Schooling has, or should have, as one of its aims the enablement of children to handle these resources and perspectives for the purpose of analysing ideas and solving problems, as indeed the cross-curricular skills imply. They are in fact *modes of understanding and experience* without which children's enablement is truncated, not parallel channels through all of which children ought to pass in order to be educated.

Therefore, they could well be grouped in a manner quite different from what we now have. No enforced caste system would be appropriate. Instead, subjects in the school curriculum could be arranged in clusters according to their relationship to descriptively different, though interacting, aspects of children's development, somewhat as shown in Figure 1.1.

In each field there should be a gradual development of the distinctive mode of thinking associated with that field, and also of a broad-brush background which would grow like a plant, however unevenly and differently for each individual. It would not, and could not, be built solely on what has served as a curriculum of subjects on a basis of custom and tradition. Nor would it involve the exiling of academic refugees.

Modes of action

So much for subjects. But within this scheme there is an element based on problem-solving and making and doing. This element ranges from

Figure 1.1 *Modes of understanding and experience*

	Languages	Mother tongue
COMMUNICATION		Other languages
	Numeracy	Mathematics
	Moral and Social Education	
·BEHAVIOUR	Religious Education	
	Practical competence	
		(Technology)
DISCOVERY	Sciences	Physical
		Biological
	Humanities	Social Sciences
		Economics
		Geography
		History
	Literatures	Mother tongue
EXPRESSION		Other
	Art	
	Music	
	Drama	
MOVEMENT	Dance	
	Physical education	Gymnastics
		Games

(Add also any number of cross-curricular links).

technology to the arts and indeed touches on all the rest, but it is not simply about understanding and experience, but about the *modes of action* that draw upon subjects.

This is where the cross-curricular elements acquire more prominence than in the present structure. As indeed the various Final Orders assiduously emphasise, these elements are closely linked with subjects, but they also embody problems and products, *doing* things rather than thinking things, though they all require thinking in order to do them. The theme with which I am most familiar, economic and industrial understanding, exemplifies this particularly well (see Chapter 5) because there is an implicit link between a mode of understanding and experience – economics, and a mode of action-industry. However, this link is not exclusive. Economics is a perspective that contributes to more than industry; industry is a mode of action that draws on more than economics.

It is interesting to notice that, in this classification, technology may figure rather as a mode of action than as a mode of understanding and experience. In that case this new subject in the National Curriculum would cease to be a subject at all, but would at once become, like personal/social education, a key cross-curricular consideration.

In a modal curriculum, such modes of action would be integral and mandatory. This might entail some proportionate limitation of subject knowledge but an increase in understanding of the significance of subjects in action. Meanwhile, the assessment of children's progress would be designed to take some account of how thoughtfully and competently they can and do perform, alone or together with others, in the various modes of action.

Modes of purpose

The third series of modes, those of *purpose*, would be rather different. They would not constitute yet another addition to the programme but a way of executing the programme; a set of permeations. They would correspond in some measure to the present 'dimensions', but they would be more than those. The essence of these modes would be that they should raise the curriculum above the level of mere instrumentality. They can provide reasons for learning and for doing. They should arise from the behavioural mode of understanding and experience, fostered through personal/social education, but they may acquire a momentum of their own as they develop into modes of being, often linked with religious or political principles. They can only develop as children develop. Like other modes, they cannot be implemented as if, instead of growing, children were delivered full-grown like Athena. As they do develop, they form a core not only of the curriculum, but also of the self. They could then constitute the basis for the consummation of secondary education through some systematic consideration of personal and social issues, rather on the lines of Stenhouse's *Humanities Curriculum Project* (1967–72), in which values and purposes as well as action, experience and understanding could be effectively mobilised.

Of course, in a democracy, purposes and values could not and should not be the same for everybody. In our time we have seen the fate of educational systems that tried to enforce the same values on everyone. Yet there are some purposes and values that *are* widely shared and necessary for the maintenance and well-being of democratic society itself. It is here that multicultural and equal opportunities issues, and awareness of other facets of social distinctions such as social class and rural–urban differences, together with positive attitudes towards different kinds of special educational needs as well as the improvement of social interaction in and out of school, can claim to be central to any modes of purpose in democratic education. Such purposes can be illuminated through the practice of the citizenship mode of action. At the same time, no one 'dimension' or its equivalent should claim primacy of importance, or exemption from the realities of children's development or from the canons of truth.

Finally, any audit of a modal curriculum would need to take account of the ways in which modes of purpose had been involved centrally in curriculum design. The fact that children could never be tested for purpose, in the way that they might be tested for action and more possibly for understanding and experience, only brings out the necessary inadequacy of any form of assessment or record-keeping that depends on simplistic testing alone.

The outcome of such an emphasis on modes of purpose should be a society of differing individuals and groups, perhaps convinced of the unique truth and rectitude of their own ideas and beliefs, but agreed on certain principles essential to a good and humane society, and practised through learning and living together well in a school community. And that should be the greatest cross-curricular consideration of all.

POSTLUDE

We do not have such a curriculum, yet. Instead, we have to contend with the National Curriculum that we do have. We cannot create our cake and eat it too. Instead, we have to try to ensure implementation of the existing elaborated agenda in a way that gives due place to the intentions of the cross-curricular elements as well as to the integrity of teachers.

This involves, first, helping teachers to be confident in their own basis of understanding and experience. That is one of the aims of the present book.

A second aim is to explore the processes of implementation themselves. The path of true innovation never runs smooth. It is important that this should be known, and action at the teachers' level taken to anticipate problems and overcome them.

However good these aims may be, it is necessary to have adequate resources with which to pursue them. Some indication will be given of how those resources can be found and deployed.

For Key Stage 4 there is the additional problem of public examinations which bid fair to make the handling of cross-curricular considerations most difficult just at the stage when, as all the *Curriculum Guidance* documents indicate, there is the greatest scope for those considerations.

Nonetheless, as the following chapters show, it is possible to achieve an impressive amount in respect of all the themes and dimensions. Yet even while that success is being achieved, it should be remembered that we could have a curriculum cake baked differently, with the cross-curricular elements used not as an icing to camouflage its defects, but rather as a spicing to add zest to its virtues.

REFERENCES

DES (1989) *National Curriculum; From Policy to Practice*, London: HMSO.

Hargreaves, DH (1991) 'Coherence and manageability: reflections on the National Curriculum and cross-curricular provision', *Curriculum Journal 2*, 1, 33–41.

HMI (1985) *The Curriculum 5–16. Curriculum Matters 2*, London: HMSO.

Miles, S and Middleton, C (1990) 'Girls' education in the balance: the ERA and equality', in Flude, M and Hammer, M (eds) *The Education Reform Act; Its Origins and Implications*, London: Falmer Press.

NCC (1989a) *Curriculum Guidance 1, A Framework for the Primary Curriculum*. York: NCC.

NCC (1989b) *Circular No. 6. The National Curriculum and Whole Curriculum Planning*, York: NCC.

NCC (1989c) *Curriculum Guidance 2. A Curriculum for All*, York: NCC.

NCC (1990) *Curriculum Guidance 3. The Whole Curriculum*, York: NCC.

Stenhouse, Lawrence (1967–72) *The Humanities Curriculum Project*, London: Heinemann.

Chapter 2

Environmental Education – An Instrument of Change?

David Wheatley

WHAT IS ALL THE FUSS ABOUT?

The existence of global environmental problems such as the green-house effect, damage to the ozone layer, the destruction of tropical rainforests and acid rain can no longer be disputed. What is open to question is the extent of the impacts each will have on the Earth and its ecology. For example, global warming may be as much as 5° Celsius over the next 40 years, which would cause serious disruptions to sea levels, plant life and the weather (Boyle and Ardill, 1989). Added to these are local problems including litter, dereliction, pollution, waste disposal and loss of habitats. The very fact that environmental problems now attract a much greater degree of international recognition is encouraging, although in many cases it seems that actions to effect solutions may not be afforded the priority their urgency demands. In the face of all these complex and controversial issues there is a danger of either becoming despondent and developing a feeling of helplessness or becoming complacent by thinking that using unleaded petrol and recycled paper is all that is needed to save the world!

Each day we all make many decisions that affect the environment. These may be from switching on a light, buying food and detergents, deciding how to get to school or work, or where to go on holiday. In the face of all the conflicting information that we are bombarded with by the media, environmental education has an important role to play in the facilitation of informed decision-making. Furthermore, it can be argued that environmental education is vital in the democratic transition to a more sustainable society, that it nurtures long-term responsibility and that comprehensive policies and programmes will save future environ-mental and economic costs through such measures as reducing the growth of energy demand and pollution levels (CEE, 1991).

Public debate and decisions, including consumer choices, require sound knowledge and awareness of environmental issues. The education system must play an important part in promoting environmental awareness, understanding and competence (HM Government, 1990).

It is worth recalling that during 1988 and early 1989 environmental education did not even appear on an unofficial list of cross-curricular themes (Dufour, 1990). This was a time when Mrs Thatcher was looking forward to her fourth term in office, the Berlin Wall separated east from west, Nelson Mandela was still in prison, apartheid was the law in South Africa and the Communist Party dominated the Soviet Union. The fact that such momentous transitions are actually taking place should suggest that anything is possible where the will and demand exists. Huckle (in Dufour, 1990) highlights the important links between environmental education, democracy, justice and the well-being of the planet, charting the development of environmentalism as a social movement and the recent growth of environmental education.

Over the last four years there has been a number of key developments encouraging a much broader interpretation of environmental education both here and abroad. International recognition of the crucial link between environmental education and sustainable development came in 1987 with the publication of The Brundtland Report (World Commission on Environmental Development, 1987) which called for a 'vast campaign of education, debate and public participation'. In May 1988 the Council of Ministers of the European Community agreed on 'the need to take concrete steps for the promotion of environmental education so that this can be intensified in a comprehensive way throughout the Community'. Following this it was resolved that member states should promote environmental education across the curriculum within all schools of the Community.

These steps were no doubt influential in the decision to include environmental education as a distinct theme in the National Curriculum. During September 1988, Mrs Thatcher gave a speech to the Royal Society backing environmental initiatives and this was followed by the publication of *Environmental Education from 5 to 16* (DES, 1989), providing guidance from HM Inspectorate. In the autumn of 1990 both the National Curriculum Council document *Curriculum Guidance 7 - Environmental Education* (NCC, 1990a) and the Government white paper *This Common Inheritance* (HM Government, 1990) were published. *Curriculum Guidance 7* provides assistance to schools in the implementation of the cross-curricular theme while the white paper sets out the Government's approach to an environmental policy and management of the environment in the UK and includes a section on education and training. The Government describe the chosen instruments of environmental policy as the use of market forces, legislation and regulation as well as incentives and disincentives.

In their response to the white paper, The Council for Environmental Education (CEE), 1991 asserts its view that, in the long term, environmental education and training is the most powerful instrument of environmental policy, reinforcing the Government's view that: 'the environment will only improve if we ourselves have the will to do what lies in our hands'.

WHAT IS ENVIRONMENTAL EDUCATION AND WHY IS IT SO IMPORTANT?

There is concern that the view of environmental education expressed in the white paper appears to be narrow, emphasising factual knowledge while not fully recognising the importance of interpretation, evaluation, awareness of pesonal values, decision-making and individual affective responses to the environment (*Times Educational Supplement*, 21 June 1991). The usefulness of environmental issues as a means for the exploration of moral, social and political values is stated by HMI (DES, 1989). The implications of these perspectives for environmental education are discussed by Huckle (see Dufour, 1990) who stresses the importance of development education in challenging environmental education to take on a more socially critical role. A plea is made for educators to take the opportunities avilable to them to consider the real roots of environmental issues allowing reflection and action on genuine social and environmental alternatives. A survey by the Global Impact Project during 1986/7 showed that the political and controversial nature of environmental education acted as a constraint in the eyes of many teachers (Greig *et al.*, 1987). However, teachers trained in the social sciences and in integrated approaches to humanities have often provided the special insights and expertise needed to support and develop such aspects of cross-curricular themes (Dufour, 1990). While social and political perspectives are mentioned in *Curriculum Guidance 7*, very little guidance is given on the provision of 'breadth and balance'. The proposed programme of study for science (DES, 1991), for instance, refers to 'the social and economic factors associated with manufacturing materials'. Environmental education must be seen as far more then just a combination of programmes of study extracted from science and geography. It also encompasses moral, cultural, spiritual, political, aesthetic and emotional dimensions. It touches on all aspects of our lives and is the responsibility of all teachers.

A definition of environmental education which is widely accepted in the UK and elsewhere comes from the IUCN (1970) Conference, which states that:

> Environmental education is the process of recognising values and clarifying concepts in order to develop skills and attitudes necessary to understand and appreciate the inter-relatedness among man, his culture and his biophysical surroundings. Environmental education also entails practice in decision making and self-formulation of a code of behaviour about issues concerning environmental quality.

Encompassed by this definition are the four overlapping components listed by HMI (DES, 1989):

a curiosity and awareness about the environment;
b knowledge and understanding;
c skills;
d informed concern.

It is informed concern on the part of individuals or the school which frequently leads to participation in activities to improve the environment.

WHERE DOES IT FIT INTO THE NATIONAL CURRICULUM?

It is apparent from the NCC Guidance and DES Guidelines that the delivery of much of the knowledge and understanding associated with environmental education should be through the core and foundation subjects, especially science, geography and technology (see case study 2, pp. 34–37). It is not always readily appreciated, but where this occurs the statutory requirements regarding assessment will apply. With the proposed slimming down of the science attainment targes, there is concern that most of the more contentious material, including global warming, pollution and radioactivity have been lost (*The Guardian*, 25 June 1991). However, the programmes of study, from which it is advised that the curriculum be developed, remain virtually unchanged and retain such important statements as 'use their scientific knowledge, weigh evidence and form balanced judgements about some of the major environmental issues facing society' (DES, 1991).

Although environmental education has been well established in the curriculum of many schools for a long time, the National Curriculum now provides a framework incorporating both an entitlement and progression for every child.

The following five models have been suggested for the implementation of the cross-curricular themes in the school timetable in *Curriculum Guidance 3* (NCC, 1990b):

- taught through subjects;
- as blocks of activities eg topics;
- as separately timetabled themes;
- taught through Personal and Social Education (PSE);
- through activity weeks or days.

The advantages and disadvantages of each approach are described (NCC 1990b; NPRA, 1989) but it is helpful to expand on these and examine the experiences of individual schools through additional case studies. Of particular value are the extensive accounts given in Neal and Palmer (1990) and Cade (1990). In the latter, attention is drawn to the importance of adopting appropriate teaching and learning strategies as well as classroom organisation. The range of approaches illustrated in the case studies consist of:

- working outside the classroom;
- bringing in outsiders;
- use of data;
- use of documents;
- expression of views through discussion or debate;
- drama, including role plays and simulations;
- projects;
- presentation.

Motivation of the pupils is a vital component of good teaching and so a valuable asset of environmental education is that such a variety of approaches can be used to generate interest and help to give relevance.

A POLICY FOR THE FUTURE

Why is it needed?

An environmental policy should be an intrinsic part of the school development plan. It serves to project the overall philosophy of the school, encompassing curriculum development and ethos. It is likely that the motivation behind the establishment of such a policy will be drawn from a sense of responsibility to the community, a response to pressures from staff and pupils, the need to save money through such acts as energy conservation or even making the school a more attractive proposition to potential pupils and their parents.

Formulation

Many local education authorities have produced guidelines for environmental education and these should be referred to at the outset. Since ownership of the policy is a critical component for success, all teachers should be involved in its formulation and implementation. Arguably, it should also involve the pupils, especially where the need for a policy has been identified by them. Concise guidance on policy formulation is given in Elcome (1991), which has been distributed to all schools in the UK. A more detailed discussion is given in Neal and Palmer (1990) who have identified eight sections for a policy statement.

There is, however, merit in keeping a policy statement concise (for example, a side of A4 paper) particularly with regard to encouraging its communication and keeping paperwork to a minimum. With this in mind, consideration should be given to the production of a series of short statements encompassing general aims, philosophy, commitment and intent with regard to the school curriculum, its buildings, grounds and facilities. It may also be of help to those involved in the formulation of the policy to consider the following:

- the resource implications needed to implement it, including staff development;
- the scope of the policy – will it extend across every facet of school life?
- any targets set should be specific, measurable and achievable;
- the support and commitment of governors, staff, pupils and parents;
- the methods and strategies adopted to communicate and implement it, including such appointments as an environmental education coordinator.

The references cited above, together with the Curriculum Guidance series will provide considerable assistance with regard to the curriculum but if the school policy is to extend beyond this then other aspects which will need to be addressed, such as:

- energy and water management;
- fabric, decor and furniture specification;
- paper and other consumables;
- cleaning processes;
- food and drink; and
- the management of school grounds.

A final point on this note is that the recent findings of the National Audit Office that nearly half of school buildings are in a crumbling and dangerous condition will be of particular significance with regard to the way pupils perceive attitudes towards the environment.

FROM THEORY INTO PRACTICE

CASE STUDY 1
PLANNING AN ENVIRONMENTAL PROJECT (PRIMARY)

The Canon Burrows Church of England Primary School is situated in a densely populated area of Ashton-under-Lyme, Greater Manchester, where a nature reserve has been created along the banks of the Taunton Brook which runs adjacent to the school. This did not happen overnight, but is a tribute to the hard work put in by staff and pupils over many years. It is an important point, often overlooked when staff are considering improving their school grounds, that the project is not a 'one-off' involving just a single cohort, but is on-going and has built-in elements of progression. To support this the school has a tradition of timetabling a slot providing first hand experience of painting, digging, path laying, litter clearance and other site management tasks. This ensures that all the pupils play an active part in the development and management of the site.

As a result of a great deal of practical experience the school has been able to produce a check list of the key stages in planning such a project (Figure 2.1). Although this pre-dates the publication of the statutory orders or technology, the process and attainment targets for this subject are easily recognisable. It is apparent, therefore, that involving the pupils at all stages will have benefits with respect to the implementation of the National Curriculum.

A critical element is the development and management plan so that future groups will also become involved and share the sense of ownership of the project. It will also help to ensure that continued use and maintenance of the site does not fall on one or two enthusiastic individuals but is the responsibility of the whole school. A danger here is that if no such plan exists the whole project will grind to a halt when the staff concerned move on.

CASE STUDY 2
INTER-DEPARTMENTAL COLLABORATION (SECONDARY)

Buckpool School is a 12–16 comprehensive in the Midlands which provides a useful example of departments working together to produce shared modules. The school's management structure had been reorganised with six 'curriculum area coordinators' replacing traditional subject heads and overseeing the teaching and learning strategies. In

Figure 2.1 *Planning an environmental project*

1. Identify need or area of interest.
2. Generate general idea for the project.
3. Check the following:
   legal aspects;
   health and safety aspects;
   ownership and permission.
4. Produce a more detailed plan.
5. Raise the support of the whole school, parents, governors and the local community.
6. Plan in full detail.
7. Consider the following resource implications:
   financing – fund raising and applications for grants;
   local expert help such as the leisure/planning department and ranger service, voluntary sector groups and other individuals;
   practical resources such as literature, tools, plants, timber, etc.
8. Implementation:
   timetabling;
   development and management plan.
9. Feedback:
   school displays, classwork, assemblies and other activities such as drama;
   communication through booklets, media and events.

Source: Mrs J Nickson, Canon Burrows CE Primary School.

addition there are five teachers who are responsible for coordinating the cross-curricular themes.

The environmental education coordinator has a clearly established job description which encompasses links with staff, pupils and outside agencies, curriculum development, general aspects and resources. The coordinator also has a concise set of goals which can be ranked according to their perceived importance. These include the establishment of an environmental education working party, liaison with feeder schools, the establishment of environmental resources and of environmental education as part of the corporate image of the school. (The role of the coordinator is seen as crucial to the success of implementation; Elcome, 1991; Neal 1991; Neal and Palmer, 1990 are recommended further reading.)

The environmental education working party contains representatives from each curriculum area. At first the group was able to identify what it was they wanted to do using *Curriculum Guidance 7* as a basic framework. It has been possible subsequently to identify and successfully implement tasks which are appropriate for use in assessment across the various curriculum areas.

As the statutory orders for geography and science between them contain a good deal of the fundamental content of environmental education, these subjects present an obvious starting point for inter-departmental cooperation. With this in mind Buckpool has developed an earth sciences course which also addresses attainment targets in technology. The course, which is taught to year 9 pupils, is based around studies undertaken on a local area known as the Green Wedge. The modules include a pollution survey of the stream and canal, the mapping of plant communities and human activities, and management of the area. On completion, each pupil is involved in presenting their work and producing displays.

The Wordsley Waterway Improvement Project (WIMPS) is a part-nership between the school, local industry, the Groundwork Trust, British Waterways and the Dudley Metropolitan Borough. The project is managed by a steering group with represention from each partner. The common aim is to utilise and improve a local stretch of canal. Curriculum materials are being produced in association with Wolver-hampton Polytechnic, practical improvement activities have been undertaken by the pupils and an environmental awards scheme has been introduced. Furthermore, the steering group has identified the importance of wider community involvement in order to ensure long-term success. So far this has involved drawing up plans for future development and presentations to canal-side industrialists. Such part-nerships provide a valuable means to promote action for the environ-ment drawing in other agencies (for example English Nature and The Learning From Landscape Trust) to assist.

Many environmental educationalists were disappointed when the statement referring to attitudes was removed from the draft environ-mental geography attainment target, especially since the non-statutory guidance, *Curriculum Guidance 7*, states that: 'Promoting positive atti-tudes to the environment is essential'.

Buckpool School has introduced an environmental awards scheme adapted from 'Understanding Science' to reinforce pupils' attitudes of concern for the environment and to encourage active participation in improvement activities. Pupils from years 8 and 9 must complete any or all of a range of 'environmental tasks' which can be 'fact finding', 'fact giving' or 'action tasks'. When a task is completed successfully the pupil is awarded a certain number of points. Certificates are awarded at each of three levels which can be kept with their Record of Achievement. The award scheme receives sponsorship enabling prizes to be pres-ented. The scheme is voluntary yet was well supported, leading to tree planting, the construction of bird boxes, rubbish clearance and information gathering.

Many other schools, including those who have taken part in the 'Environmental Enterprise Award Scheme' have entered their pupils for national award schemes such as the CREST Award (Creativity in Science and Technology). Schemes such as this not only act as an incentive to pupils but also provide a profile of skills achieved.

CASE STUDY 3

THE SUSPENDED TIMETABLE

Special events such as industry or environment days and weeks are becoming more and more widespread as a means of providing inter-departmental collaboration. There are benefits and problems associated with such events (NCC, 1990b; NPRA, 1989). Of these, of great importance is the increased motivation through doing something novel. Cooperation between the staff is best achieved when the activities can be related to the core and foundation subjects and successful events can often lead to further collaborative ventures developing. Unfortunately, special events are very often regarded as the solution to the provision of cross-curricular themes but are best viewed as an addition to the day-to-day planned provision of the themes.

Environmental education and economic and industrial understanding

'. . . the environment is increasingly an industry in its own right, and shows how important environmental improvement can be to economic regeneration'.

(HM Government, 1990).

A casual glance at the NCC Curriculum Guidance documents will reveal the interdependency of all the cross-curricular themes but in particular much common ground exists between economic and indus-trial understanding and environmental education. This is not surprising since trends in recent years seem to indicate that the environment has become, or soon will become, a major factor in determining change in business. This situation can be recognised in government policy and legislation such as that proposed in the White Paper, *This Common Inheritance* (HM Government, 1990) and the 1990 Environmental Protection Act (or Green Bill). Indeed, the present government is committed to test the use of market forces in promoting environmental protection and regeneration and many businesses are already seizing the commercial opportunities offered by waste reduction, energy conservation and the rise of 'green consumerism'. It is clear that the impact of environmental issues on the business world will continue to be considerable (Elkington *et al.*, 1988).

Curriculum Guidance 4 (NCC, 1990c) goes some way towards identifying the importance of the environment in economic and industrial activity:

Appreciate some of the environmental and social issues associated with economic and industrial activity.

Economic activity and growth creates national and international wealth, but can use up finite resources and damage the environment.

What really made the 'Environment Day' staged by Knowsley Hey School on Merseyside 'special' was that they made good use of their many industrial contacts to help run the activities. The day was timetabled into four one-hour sessions with a plenary session and display. In all, 15 different activities were available for the pupils of year 10 to choose from. These included paper recycling, soil analysis, making nesting boxes, water pollution, energy loss, open cast mining and waste-derived fuels with session leaders from local industry. Each activity was identified with appropriate parts of the curriculum from science, mathematics, geography, technology and English. Industrialists experienced education at first hand and this further strengthened links both with industry and with a feeder primary school which was also invited to take part in the day. Not only was the whole day reported by the public for inclusion in their in-house publication, but it also received local press coverage.

Curriculum support – the voluntary sector

A considerable amount of resources such as educational materials and facilities for teachers comes from the business and voluntary sectors (CEE, 1991). In addition, corporate sponsorship is increasingly a source of income for some voluntary sector organisations (Elkington *et al.*, 1988).

Many of the voluntary sector organisations are already well known to schools, such as the Royal Society for the Protection of Birds, the British Trust for Conservation Volunteers (BTCV) and the World Wide Fund for Nature. The resources which these organisations are able to provide are generally well publicised and accessible. In addition, voluntary sector networks can provide valuable local assistance and represent the interests of a wide range of organisations.

CASE STUDY 4
PARTNERSHIP FOR ACTION

An illustration of a network's involvement with curriculum activity is that of the Mersey Basin Trust. This network has produced a range of educational matertials which have been widely used by schools and are now incorporated into the 'Stream Care' project. In this project the network acts as a facilitator to bring together many different facets of the community with the common aim of developing long-term concern for, and active participation in, the improvement of their local watercourse.

In Kirkby, on Merseyside, two schools have taken part in a cross-phase project involving the local brook, a tributary of the River Alt. Pupils from Cherryfield Primary and Brookfield Comprehensive schools were introduced to the brook through investigations developed by the network. This involved aspects from National Curriculum science, geography and technology and included the monitoring of freshwater life, measurement and recording of pollution levels, research into the area's history, land form and land use. The pupils were also encouraged to record their own perceptions of the area and they were later able to enter all the information into a computer database which allowed them to find out more about the area and compare it with other sites in the region. Following these visits, the pupils have been involved in waterside planting with the help of the BTCV. Both schools then cooperated in the design and production of a trail leaflet which was distributed throughout the local community and an exhibition of the project was mounted in the town. For the primary school children, there was a visit to a local printing works where they were able to experience the production process first hand and print their own posters and booklets.

POST-16 EDUCATION

The interdisciplinary nature of environmental education makes it an ideal context for a planned scheme of core skills activities such as the 16–18 Curriculum Enrichment Programme, developed at the University of Liverpool. This involves students embarking on problem-solving projects which encourage them to seek advice from external sources in both industry and higher education. Projects which have been completed successfully include energy surveys, land use studies and acid rain investigations.

> **Environmental education and careers education**
>
> 'Improving the environment calls for a better trained and more aware workforce and will provide new job opportunities'.
> '. . . the Government will ensure that vocational standards are in place across the whole range of environmental jobs including countryside work, urban conservation and all aspects of pollution control and waste management'.
>
> (HM Government, 1990).

RECENT TRENDS

Life-cycle analysis

Life-cycle analysis (LCA) is a rapidly emerging technique which focuses attention on the environmental impact of products right across the life cycle. It does this through attempting to quantify all material and energy inputs and outputs through industrial systems (Figure 2.2). Early development has been carried out for the packaging industry, including milk and fabric conditioner packaging and carrier bags (ENDS, 1990, 1991). In addition, the motor car industry is carrying out analyses of a similar nature (*The Independent on Sunday*, 14 July 1991).

Packaging provides a context which is used widely by schools for industrial simulations in design and for introducing issues such as waste and recycling. In view of this, the availability of reliable information of this type will provide a valuable resource for exercises in evaluation and decision-making.

The imminent introduction of eco-labelling schemes both in the

Figure 2.2 *'Cradle to grave' steps in production*

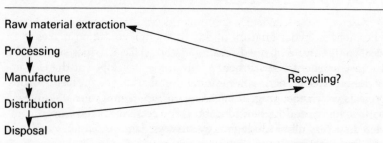

United Kingdom and the European Community, coupled with the notion that there is no such thing as a 'green product', will ensure continued refinement of this technique.

Environmental education and health education

'. . . the quality of the air, land, rivers and coastal waters of our densely populated country is still under pressure from the very economic development which improves the quality of our lives in so many ways.'

'. . . good quality air is essential for human health and the health of the environment as a whole.'

'It is vital to protect the quality of drinking water which has to meet stringent standards to safeguard health.'

(HM Government, 1990).

Environmental audits

A number of local authorities have been involved in carrying out audits to discover the impact on the environment of all aspects of their operations. Environmental audits have been used with success as the subject for cross-curricular activities in schools, often with the additional benefit of saving the schools a substantial amount of money. Schools in Lancashire have reportedly saved £450,000 in an energy-conservation scheme operated by the Lancashire County Council. As this represents a reduction in atmospheric discharges of 47,000 tons of carbon dioxide and 377 tons of sulphur dioxide over three years the significance of such a scheme cannot be overstated.

IMPLEMENTATION – A SUMMARY

- Appoint cross-curricular coordinators to develop policy in school and assist feeder schools.
- Win the support and cooperation of the senior management team, governors, staff and pupils.
- Keep paper work to a minimum.
- Involve outside agencies, including the local community, voluntary sector organisations and industry, and form partnerships.
- Use a third party to serve as a facilitator.
- Limit your initiatives to what is achievable.

Environmental education and citizenship

'We set out in this White Paper how everyone can help and what everyone can do . . . so that caring for the environment becomes an instinctive characteristic of good citizenship.'

(HM Government, 1990).

In reading this we should be reminded that all of us, politicians, industrialists and students, are citizens of the global community.

REFERENCES

Boyle, S and Ardill, J (1989) *The Greenhouse Effect*, London: Hodder and Stoughton.

Cade, A (ed.) (1990) *Opening Doors For Science*, York: ASE/NCC.

Council for Environmental Education (CEE) (1991) *Beyond This Common Inheritance: Education and Training*, Reading: Council for Environmental Education.

DES (1989) *Environmental Education from 5 to 16*, London: HMSO.

DES (1991) *Science for Ages 5 to 16*, London: HMSO.

Dufour, B (ed.) (1990) *A Guide to Cross-curricular Issues*, Cambridge, Cambridge University Press.

Elcome, D M (1991) *Environmental Education: The Vital Link*, Sandy: RSPB.

Elkington, J, Burke, T and Hailes, J (1988) *Green Pages*, London: Routledge.

ENDS (1990) *Report 188*, September, London: Environmental Data Services Ltd.

ENDS (1991) *Report 192*, January 1991, London: Environmental Data Services Ltd.

HM Government (1990) *This Common Inheritance – Britain's Environmental Strategy*, London: HMSO.

International Union for the Conservation of Nature and Natural Resources (IUCN) (1970) *Final Report, International Working Meeting on Environmental Education in the School Curriculum*, IUCN.

National Curriculum Council (1990a) *Curriculum Guidance 7. Environmental Education*, York: NCC.

National Curriculum Council (1990b) *Curriculum Guidance 2. The Whole Curriculum*, York: NCC.

National Curriculum Council (1990c) *Curriculum Guidance 4. Economic and Industrial Understanding*, York: NCC.

Neal, P (1991) 'The implementation of environmental education as a cross-curricular theme in secondary school, *Environmental Education*, 36.

Neal, P and Palmer, J (1990) *Environmental Education in the Primary School*, Oxford: Blackwell.

NPRA (1989) *Cross-curricular Approaches in Schools (1988–89) – An NPRA Investigation*, Northern Partnership for Records of Achievement, TVEI, Manchester: JMB.

World Commission on Environmental Development (1987) *Our Common Future.* The Brundtland Report, Oxford: Oxford University Press.

Chapter 3

Health Education in the National Curriculum – 'All Change' or 'As You Were'?

Pat King

INTRODUCTION

Of all the cross-curricular themes, health education may be the one most often taught in schools as a distinctly recognisable topic. This does not mean that it is unnecessary to consider it in a new light. The nature and aims of health education in a school may not have been thought through, it may not be planned and structured, and staff may be hazy and unsure of appropriate teaching methods and strategies for delivery. The purpose of this chapter is to examine some of the reasons for health education, its distinct nature as a cross-curricular theme, some of the problems it presents for teachers and strategies for coping with them. It is not possible here to provide comprehensive answers for those responsible for coordinating, planning or delivering health education as a cross-curricular theme. The intention is to pose questions, raise issues and where appropriate point towards possible ways forward. The discussion and exploration required for a school or individual to move on from there is an essential aspect of developing an approach appropriate to local needs.

WHY HEALTH EDUCATION?

At its simplest and most naïve level, health education is about avoiding ill-health and illness. In order to do this, some knowledge of the causes of illness is helpful along with some 'dos and don'ts' for individuals and communities. Yet health is a positive state, not merely the absence of illness. An understanding of the various aspects of being healthy, the factors and behaviours which affect health, and strategies for maximising one's health potential both long and short term are intrinsic elements of a positive approach to health education. It is a truism that health is one's most valuable asset, and its protection should therefore be a high priority. It is increasingly evident that attitudes and patterns of behaviour established in childhood and adolescence can have long-term influence on adult health. Thus the education of the young to develop an awareness of healthy lifestyles is vital. This is recognised by Government in its support for the Health Education Authority (HEA), by the World Health Organisation (WHO) in many aspects of its activities, and by the Department of Education and Science (DES) in its inclusion of health education as one of the cross-curricular themes which form an essential part of the whole curriculum.

A question which does not appear to be clearly resolved is that of where responsibility for health education should lie. The complex blend of facts, attitudes, skills and behaviour means that there is no definitive answer, and responsibility is shared between various agents, principally parents, health professionals and schools. It is this dilemma about where responsibility should lie which may cause considerable anxiety for teachers.

If parents are to have sole responsibility, they can transmit cultural norms, practices and values as they see fit. The success of a model which gives responsibility entirely to parents will, however, depend on satisfactory levels of factual knowledge in the population at large, as well as the ability of parents to discuss personal issues in which their own behaviour is not health promoting. It puts the onus on parents to develop their children's personal and social skills and perceptions, and in so doing assumes that all parents are able to do this. The danger of the transmission of existing levels of ignorance, prejudice and bad practice is evident. Yet parents have rights concerning the values inculcated in their children and awareness of this can make teachers wary of parental disapproval.

Health professionals claim a role in the health education of children, and have historically been used as visitors to 'do' certain delicate, controversial or sensitive topics in schools, particularly in such areas as sex and drugs. Health authorities employ personnel with a specific health promotion/health education brief which encompasses work in schools. As Burrage (1990) points out, the perspectives of these officers arise from medical, health authority origins. Health education, as part of personal and social education (PSE), is a broader topic than this perspective may allow, and whatever their health-related expertise, such officers often lack educational perspectives and skills. The out-dated view of health education as a factual area designed to give information about appropriate behaviours was linked to the use of outside 'experts' who hold knowledge not available to teachers. The current broader view needs a re-assessment of this approach. In addition, the HEA's national publicity, promotion and education campaigns can complement work done in schools.

The other group who may be held responsible for health education is teachers. Few teachers are trained to deliver health education, just as few are trained in PSE. Those who take some training are almost always principally concerned with another specific area of the curriculum. Teachers may be called upon to deliver health education without training and their lack of factual knowledge, together with an awareness of the values, moral issues and cultural elements involved may make them reluctant to take the area on board. In health education, many sensitive issues have personal elements which make them difficult to discuss with others. If health education is to be part of the school curriculum however, it should not be divorced from other elements and cannot be handed over entirely to health experts or parents.

There is thus no one group who can claim total ownership of health education. All three, health staff, parents and teachers have something to offer, and a cooperative approach between all parties is likely to result in the most productive strategy. This requires an exchange of views and discussion of issues for which opportunities need to be created.

HEALTH EDUCATION, PAST AND PRESENT

Health education has a long history if we consider general 'education for health' rather than purely school-based work. As medical knowledge grew and offered explanations for the cause of ill-health, so the populace became aware of these and strategies for avoiding illness. The Health Education Authority's training manual, *Health Education 5–13* (HEA, 1983), suggests a consideration of causes of ill-health at the turn of the century and comparison with those of today, at the end of the twentieth century. The change in the developed world from transmissible diseases and the effects of deprivation, to our present-day ills, many caused by lifestyles and affluence, reveals the different nature of health education needs. The earlier emphasis on hygiene, diet and 'teeth, nits and naughty bits' is now inappropriate. Many studies have revealed the lack of influence of extra knowledge on individual behaviour. It has also been shown that work which links attitudes and skills with knowledge has measurable effects on future behaviour (Whitehead, 1989). As Tones *et al.* (1990) point out, knowledge is necessary but not sufficient for effective health education. Many people do have choices, albeit limited by social and political factors. Health education thus needs to address the issue of choice and influences, and empowerment to give people social and political effectiveness. There is a danger of prescription if issues are not adequately explored; there is usually a wide range of choice in healthy behaviour (Dalzell-Ward, 1974).

The aspects of the health education curriculum which seem to be regarded as important are those which are 'panic issues', such as sex, drugs and AIDS, topics which cause fear in parents of teenagers. This problem-centred approach to health education may be forced on schools by public opinion. Yet at the same time there is concern over these issues in many quarters, over factual knowledge encouraging experimentation, or the avoidance of discussing personal elements of sexuality because of legal considerations and long-standing taboos. The other cross-curricular issues do not suffer from quite the same public hang-ups.

An emphasis on choice, attitudes, influences, assertiveness, decisions and consequences to accompany facts reasserts the place of health education as part of PSE and reveals the links between the cross-curricular issues since these are fundamental to all of them.

Whitehead (1989) suggests three key principles for health education which would be a firm foundation for future development and carry us from present good practice into the twenty-first century. The essence of Whitehead's suggestions is:

- Children need help in clarifying their own ideas and attitudes. As well as information, they need to practise making choices. Active learning provides opportunities for this.
- There are clear links between health behaviour and self-esteem. Methods of working which build self-esteem are therefore essential.
- Isolated lessons on specific topics are not enough. Key themes need to be revisited to match increasing maturity. A spiral curriculum needs to be carefully planned.

These points need to be borne in mind when interpreting any documentation concerned with health education.

THE DOCUMENTATION

Curriculum Guidance 5 (NCC, 1990) from the National Curriculum Council is not the first advisory document on health education. The HMI 'Curriculum Matters' series produced a useful comprehensive overview of health education (DES, 1986). The HEA and Schools Council have published guidance for this area and much has been published about specific areas such as sex education and drugs. The attention given to *Curriculum Guidance 5* (NCC, 1990) stems from its origins in the NCC as part of the overall documentation gradually providing authoritative parameters for the curriculum.

As health education is not new, much of the document presents few surprises. It is clearly set out and offers nine areas of health education with suggestions for each key stage, facilitating the establishment of a developing and progressive curriculum. However, the brevity of the guidance has resulted in many simple global statements for what are complex, multi-faceted topics. There is ample scope therefore for individual interpretation, which is a double-edged sword. Development may be minimal with a very cursory interpretation, which would not be helped by the percentage of statements based on knowledge and not spelling out other relevant aspects. Alternatively, the brevity of the information could be a great advantage, since it allows schools to make their own interpretation of the guidelines. This process of articulating the meaning of the statements, linked to the needs of the pupils, the school's aims and ethos and local guidelines, is valuable in-service training for staff involved and would ideally be shared by a team. In spelling out the meanings of the simple statements in the guidance document, curriculum links are revealed, facilitating planning; teachers' own values and attitudes are debated; awareness is raised and overlaps while other cross-curricular areas emerge. Knowledge, attitudes and skills can be identified and appropriate strategies for delivery discussed. Thus the apparent deficiency in the documentation, with inadequately developed simple statements, is in fact an advantage because it allows staff and curriculum development through the process of clarifying details in the light of an individual school's needs. This is only true, however, where an appropriate amount of time and effort is made available, that is, where health education is given its deserved significance.

The existing state of staff awareness can be a limiting factor in the successful development of health education. Teachers may still be holding an authoritarian, dogmatic view of health education, with a preaching style. As Dalzell-Ward says, the aim is not to bully into health but liberate from unnecessary fear and anxiety, and help develop the utmost potential. This positive view needs to be spread universally if the holistic, lifestyle-centred approach is to succeed and if the documentation is to be sufficiently developed to produce a successful individually-tailored health education programme.

Going through the process outlined above could be helpful in raising awareness and moving teachers' positions on the nature and purpose of health education.

STRATEGIES FOR IMPLEMENTATION

Our current understanding of health education reflects the growing awareness of the various factors which influence behaviour. The past emphasis on factual knowledge as a key to altering personal behaviour of individuals is now seen as deficient. The attitudes and values of individuals and society are fundamental, as is the possession of skills to cope with health-related aspects of life. Health education must therefore aim to develop all three elements. There are implications here for teaching style. Facts can be transmitted by traditional didactic methods and attainment tested objectively. Values and attitudes, if they are to be developed rather than passed on by indoctrination, require open discussion and exploration, while skills need to be practised if competence is to be increased. Much has been written on the teaching styles appropriate to PSE, emphasising the need for active learning and the importance of the process as well as content. Sufficient to point out here that if health education is to be integrated across the core and foundation subjects, all teachers need to be competent in the participatory teaching styles involved, such as brainstorming, discussion, use of trigger statements or pictures, role play or drama. Moreover, they need to be prepared to deliver the aspects of their subject which encompass health education in ways which are compatible with good health education. For many this will require a dramatic shift in position, although there is already a trend towards using a variety of teaching methods in many subjects. The importance of articulating clearly the appropriate aims, objectives and content of health education and identifying knowledge, attitudes and skills involved is evident, to allow appropriate teaching styles to be selected. There is a danger that a global statement will be seen to suffice with little consideration of the implications for detailed content and appropriate learning experiences.

An example of such specific articulation of objectives can be seen if we consider sex education. A simple statement on this topic, in *Curriculum Guidance 5* (NCC, 1990, p. 14), referring to Key Stage 2, is that children should 'begin to know about and have some understanding of the physical, emotional and social changes which take place at puberty'. How can we help children to understand rather than merely 'know'? What exactly is relevant to this age group? We may include feelings of being different, of not knowing what to expect or what is expected, feelings towards friends of each sex, feelings of attraction, of excitement, of sexual interest, about pressures to conform from the media, peers or family. More importantly, we may include how to deal with such feelings. Although time-consuming, this approach, of fleshing out general statements, allows easier checking of coverage and identification of relevant teaching strategies. Sex education may, for example, be assumed to be covered in science as part of the National Curriculum statutes. Looking at the expansion of one simple statement above it is apparent that only a fraction would be relevant to science, or indeed to any one subject. By spelling out clearly what is entailed, planning for complete coverage is facilitated.

The notion of a cross-curricular theme suggests a spread across a number of subject areas. The NCC documentation is at pains to point out, however, that there are various possible models for delivery, which may be across subjects, in discrete time, a combination of these or in other ways. Unlike the cross-curricular dimensions there is no expectation that the themes will permeate the whole curriculum, yet there are some aspects of health education which need to permeate the school ethos. Other aspects of school life, such as work in assemblies, visits, the way people speak to each other and the amount of responsibility or restriction on pupils, may be transmitting messages and attitudes which are contradictory to those planned in the formal curriculum. The distinction between themes and dimensions is therefore less clear-cut than at first appears. It is useful to view health education as a theme when considering the formal, planned curriculum with an accompanying examination of the life of the whole school to ensure that the dimensional aspects are consistent and complementary.

Delivering health education through a cross-curricular approach is initially attractive. It appears to save time, to ensure whole-school involvement and make subjects more relevant to everyday life. The first reaction of schools to the requirements of the National Curriculum included some shock and trepidation at the prospect of fitting ten subjects into the timetable for all pupils. The idea of removing separate time in secondary schools for PSE or health education, and assimilating them into the core and foundation subjects seemed to be a helpful way forward. Some elements of health education are already addressed in subjects as a matter of course as part of the National Curriculum requirements. Others could be introduced by a change of emphasis, careful selection of examples or by some cooperative work between departments.

The disadvantages of such an approach become evident on closer consideration. All teaching staff need to be aware of the aims of health education and be prepared to adopt appropriate teaching approaches for the various aspects. The limited number of staff already trained to deliver health education suggests an enormous in-service training programme would be needed. This would also be needed for the other themes if a totally cross-curricular approach were to be adopted.

National Curriculum and examination requirements for all subjects already put pressure on time. When time is short it would be a very dedicated health educator who put completion of the syllabus at risk to give due emphasis to health education topics. This is not a cynical view of teachers as much as a realistic acceptance of the pressures they are under. At the end of the day they are often judged by the examination success of their pupils rather than by the adequacy of the health education covered.

Coverage of parts of a topic by separate subjects also leads to fragmentation, with pupils being expected to put related ideas and concepts together themselves. This transference may well not happen.

What of the obvious alternative, a timetabled slot to deliver the themes, perhaps entitled PSE? This takes up precious time and therefore needs careful justification. The advantages include a cohesive approach, avoiding fragmentation, and the possibility of having a team of staff trained in particular skills. Pupils are not expected to transfer ideas between subjects and can explore different aspects of a topic together. Timing problems, where topics could be addressed in different subjects months apart, are similarly avoided. This approach can, however, be problematic if seen as a total solution. Health education may duplicate some elements of work done in subjects, wasting time, causing boredom and disaffection through repetition, and possibly giving conflicting messages. Subject teachers may be encouraged to consider health education as none of their concern, losing opportunities for enhancing their own teaching and for developing the dimensional aspects of health education.

A combination of these two approaches, possibly with other activities, seems a useful way forward. It offers the advantages of both with a minimisation of the disadvantages. The disadvantage that it introduces is the need for even more careful coordination and cooperation. The role of the PSE or health education coordinator would be much more demanding and require commensurate recognition and status.

Just as each school makes its own interpretation of the other National Curriculum documents and delivers them in a way which is consistent with its own values and priorities, so health education will be delivered in an individual way in each school. This is likely to be a unique amalgam of subject-based work, separate provision and special events. Whatever model is selected, it will require careful planning of what is to be addressed for each year group if adequate monitoring and evaluation are to be possible.

EXAMPLES OF APPROACHES

Primary and secondary schools traditionally have a fundamentally different approach to the curriculum. Primary education has a reputation for 'topics' used to capture children's interest and deliver the various assessed subjects of the curriculum. On the transition to secondary school pupils are taught separate subjects, and the inter-relationships developed and exploited in the 'topic' approach are lost. Elements of health education, as with the other cross-curricular themes, often form the basis of a topic in primary schools and this holistic approach has its advantages, but it does not lend itself to the usual secondary school timetable organisation and departmental structure. A re-examination of the possibilities for this could be fruitful.

A topic such as 'The Family' could involve work with several departments, especially English, RE, geography, history and mathematics, plus PSE, and deliver health education, citizenship, equal opportunities and careers materials. It would avoid fragmentation and mistiming as aspects would be planned together and delivered at the same time. A topic could last for half a term or less, with joint inter-departmental planning to give coherence and ensure that the requirements of each subject were met. Aims and objectives would need to be clearly set out, not only for subjects but for the cross-curricular elements.

As schools are re-examining their curricular provision in the light of National Curriculum requirements, fresh approaches to health education and PSE are being tried out. For example, in an 11–16 comprehensive school with a well-structured timetabled PSE programme, a 'Health Day' was organised for all pupils in year 8. The timetable was suspended for the day for this year group. Ten outside specialists and volunteer teachers devised one-hour activities (timed to match standard lesson length to confine movement of the pupils to normal times) to highlight specific topics in the year 8 health education curriculum. Pupils were placed in ten groups and each group attended five different activities. Pupils recorded their views and responses and shared these with classmates in subsequent PSE lessons. The value of this was that it allowed for specialist input without excessive calls on the time of the 'experts', and without the visitors being seen to 'do' the topics as they were also included as part of the normal health education curriculum. It also raised the profile of health education in the school. Teachers accompanied the pupils at the times when they would normally have been teaching this year group, so some staff gained insights into aspects of the year 8 health education materials which they would otherwise not experience. The success of this approach depended on long-term planning (six months) and minimisation of disruption to the normal daily life of the rest of the school.

In another 11–16 school, where health education was being reviewed and developed from a low base-line, a cross-curricular approach was used for a particular health issue – smoking – to coincide with National No Smoking Day. Staff were presented with statistics about local patterns of smoking-related health problems, and all departments were invited to take part, with no coercion; most did so, devising activities to deliver their subject using the theme of smoking. All this work was timed to be in the same week. The success here depended on the lack of pressure to take part and the preparation through presenting relevant information and linking to a national event.

Neither of these ventures could be a model for the whole health education curriculum but add variety and colour to the basic arrangements.

In primary schools, health education remains a popular source for topic work. In planning other topics, however, it is fruitful to note the scope for delivering aspects of the themes (a simple 'CHEEC' check – the initials of the five cross-curricular themes) at the same time as planning the maths, geography, English, etc. to be covered by the topic. When specifically teaching other subjects it is also possible to use health education examples, but the health message is often partial and may be lost. These strategies alone would lead to a patchy approach, with fragments of health education covered when convenient for other topics, so an overall view of what is desired and where it is happening is also necessary so that omissions can be dealt with in future, or amended, planning.

Where health education is the basis of a topic the principles of a curriculum which revisits topics as pupils' experience and maturity grows (a spiral curriculum) needs to be borne in mind. Thus food cannot be 'done' in year 2 and year 6, for example. Different aspects need to be addressed with each age group. The approach in one primary school was to have a theme of 'Healthy Eating' for a half-term topic for each year group. Teachers selected aspects of this theme relevant to their class and the culmination of the half-term's work was a healthy eating week with visitors, parents and sponsors involved. This was planned in conjunction with the LEA school catering service. This approach allows for the re-use of the planning as a basis for the following year, as pupils have not covered the same aspects of food earlier in the school.

Another approach is the 'health week' where all work in a particular week is on an aspect of health education, different for each year group, eg, smoking for year 6, food for year 5, hygiene for year 4, etc. This allows a high-profile approach with visitors and parental involvement, but there is a danger of it being used to replace an on-going planned coordinated programme to provide a spiral curriculum.

Yet another strategy that has been devised is to take the nine areas of, for example, Key Stage 2, decide in detail on the important aspects of each area, then divide them into 12 or 24 'topics', each to cover a term or half a term in the four years of Key Stage 2. These are allocated to age groups seen as appropriate in that school and the class teacher, with the health education coordinator, plans the term or half-term topic. This overcomes the haphazard arrangements which can occur when class teachers choose their own topics, and allows for major areas to be split and different aspects to be addressed with different age groups. It also facilitates tracking of the health education programme.

These few examples show the scope for individual interpretations. Each school does, however, need a coordinator for PSE or health education to develop and organise the best combination of strategies and activities for the school's individual circumstances.

INTERWOVEN NATURE OF THE THEMES

Curriculum Guidance 3 (1989), the precursor to the five 'theme' documents from the NCC, evades the issue of the relationship between PSE and the cross-curricular issues. This uncertain relationship can cause tensions in schools between those responsible for and experienced with PSE and those involved in introducing cross-curricular issues. A sensible approach is to merge the two responsibilities; the two are not synonymous but have so much in common that working separately can be a nonsense. The development of personal and social skills, exploration of attitudes and raising self-esteem are integral to all the cross-curricular issues as well as to PSE.

There may be some value in taking an holistic view of all cross-curricular themes rather than planning for each of them separately. This avoids inevitable overlap, evident in the guidance but even more apparent when the statements and suggestions in the documentation are fleshed out. Work on the family is an obvious instance, being almost duplicated in health education and citizenship. Education about the community is less obvious, but has facets in health education, citizenship and the environment. Such overlaps abound and emphasise the arbitrary divisions into distinct themes. It could be economical in time and energy in the long term, and more coherent for pupils, if the themes are merged for practical purposes, producing new amalgamated topics which draw together elements from various themes. If objectives are clearly articulated tracking of individual themes should be straightforward. This is not the place to discuss the idea further but it is worth considering when planning for cross-curricular issues in the curriculum.

PLANNING FOR HEALTH EDUCATION

The role of the health education coordinator (which may be combined with that of PSE or cross-curricular coordinator) is demanding and challenging. For success, the school organisation should recognise this. The post-holder needs vision, confidence in active learning, the ability to articulate ideas, communication skills, an organised mind and boundless energy; a thick skin may also be an advantage! A 'health' background may not be essential, and teachers from any subject area can succeed if they acknowledge this. The medical elements of health-related knowledge at school level are easily acquired and there are many experts in the community whose help can be enlisted when necessary. The generic skills of a good coordinator are more important than specific health knowledge gained from science, PE or home economics teaching, traditional curriculum areas associated with health.

A suggestion for planning

A starting point for a coordinator could be to look at the *health entitlement* of the pupils in the school. This may be informed by DES, NCC and HEA guidance, by surveys such as those from Exeter University (eg, Balding (nd)), by discussion with parents, governors, health authorities and LEA staff, or other sources. Health education could be the focus of work of a team in the school, whose brief would be to determine what is *appropriate* for the pupils at *each stage* through the school.

- The statements of entitlement need to be *clearly articulated*, ideally as objectives, and spell out relevant *knowledge*, *attitudes*, and *skills* involved in each area.
- After the pupils' health entitlement has been formulated, *checks* should be carried out to ensure that the topics identified in *Curriculum Guidance 5* have been addressed appropriate to each key stage.
- *Governors* could be presented with the document at this stage for comment. (Many governors are surprised at the scope and complexity of today's health education).
- In *primary schools*, staff can be asked how their *plans* for the class's work over the year encompass the identified health education material.

- In secondary schools, *heads of department* can be briefed (ideally departmental representatives would be involved in the planning team) and asked to *identify health education* as articulated in the school's documentation within their subject. Coverage in PSE/ health education lessons could also be audited. The coordinator would have a mammoth task in charting this, but *omissions* and *duplications* would be identified.
- The coordinator and planning team could then *negotiate* with heads of departments (class teachers in primary schools) to encourage cooperative working in providing *coherent coverage*.

This outline strategy is obviously sketchy but space prevents more detailed suggestions. This general approach has been tried and tested and can be used for other cross-curricular themes, or a PSE course encompassing all the themes and dimensions. This of course increases the complexity and makes the task more daunting. It is not easy to do (although relatively easy to describe!), and takes time, persistence and determination. This is why the coordinating skills are more important than the health knowledge for the person designated to do this work.

CONCLUSION

In the light of the above, the questions posed in the title can perhaps be addressed. As is typical in health education, there is no single answer. For those who have developed health education in their school thoroughly and comprehensively, especially in a PSE and/or cross-curricular context, it will be mostly 'as you were', since there is little controversially new in the NCC Guidance. Some refinement may be necessary, with a change of emphasis, additional topics or a change of timing. In particular, some aspects may be introduced earlier than has been common practice in the past. A well coordinated, successful health education programme may well need little alteration.

For many schools it will indeed be 'all change' since it is common for health education to be somewhat ad hoc, patchy and disjointed, despite good intentions. Here it is to be hoped that the NCC documentation will give the impetus for a review of health education, preferably integrated with PSE and the other cross-curricular issues, with a commensurate rise in status and a determination to develop a coordinated, comprehensive spiral programme to provide the health education entitlement of the pupils.

REFERENCES

Balding, J (nd) *Just a Tick*, Exeter: Schools Health Education Unit, Exeter University.

Burrage, H (1990), 'Health education: education for health?', in Dufour, B (ed.) (1990) *The New Social Curriculum: A Guide to Cross-curricular Issues*, Cambridge: Cambridge University Press.

Dalzell-Ward, AJ (1974) *A Textbook of Health Education*, London: Tavistock Publications.

DES (1986) *Curriculum Matters 6. Health Education from 5–16*, London: HMSO.

Health Education 13–18, Developing Health Education: A Coordinator's Guide, Schools Council/HEC Project, London: Forbes Publications.

HEA (1983) Schools Health Education project 5–13, Training Programme, Health Education Council.

Curriculum Guidance 5. Health Education, York: National Curriculum Council.

Tones, K, Tilford, S and Robinson, Y (1990) *Health Education, Effectiveness and Efficiency*, London: Chapman and Hall.

Whitehead, M (1989) *Swimming Upstream: Trends and Prospects in Education for Health*, Research Report 5, London: King's Fund Institute.

Chapter 4

Careers Education and Guidance across the Curriculum – Making it Work for You

Jim Burke

INTRODUCTION

This chapter attempts to provide teachers in primary and secondary schools with practical suggestions relating to the implementation of careers education and guidance across the curriculum. It will deal with the basic questions of:

- What is careers education and guidance?
- Why do schools need to provide a careers education and guidance programme?
- What kind of programme should be provided?
- Where do schools begin when planning and implementing such a programme?

Two case studies are provided which indicate how a primary school and secondary school have attempted to implement careers education and guidance across the curriculum, working from policy development through to practical implementation. Examples of the documentation produced by the schools during their development phase is also provided.

WHAT IS CAREERS EDUCATION AND GUIDANCE?

Careers education and guidance (CEG) is one of the five cross-curricular themes described by the National Curriculum Council (NCC, 1990) as promoting the aims defined in Section 1 of the Education

Reform Act 1988, ie, 'to prepare pupils for the opportunities, responsibilities and experiences of adult life'.

CEG aims to help pupils develop skills, attitudes, knowledge and abilities which will prepare them for coping with the various stages of their lives. Its fundamental concern is to ensure that all pupils are able to make informed choices as they progress through school and beyond into adult life.

CEG is a process which must commence in the primary school where so much can be done to lay the foundations for developing the skills and knowledge necessary to make effective choices. Furthermore, from an early age all pupils should be helped to guard against sex and racial stereotyping as these can constrain and inhibit subsequent choices and aspirations.

Careers is a term used to describe the variety of occupational roles which individuals will undertake throughout their life and includes paid and self employment; different occupations which a person may have during adult life; periods of unemployment; and unpaid occupations such as that of student, voluntary worker or parent. *Guidance* is a term used to describe the processes which help individuals to make choices and transitions appropriate to their needs and aspirations.

WHY DO SCHOOLS NEED TO PROVIDE CEG?

Traditionally, of course, there has been little pressure on schools to provide CEG and in the majority of secondary institutions pre-1980, careers departments were generally held in very low esteem or were simply non-existent.

Since the mid-1980s, however, all schools, including primary, have been under steadily increasing pressure to seriously address the question of CEG provision. *Better Schools* (DES, 1985) stated, 'A major objective of education is to help pupils acquire understanding, knowledge and skills relevant to adult life and employment in a fast changing world'. The Technical and Vocational Education Initiative (TVEI) (1983) made the provision of careers education a contractual commitment for all participating schools and colleges and undoubtedly raised the status of CEG as well as providing much needed resources.

'Working Together for a Better Future' (DES/DoE, 1987) initiated a review of CEG at both LEA and individual school level and reiterated the entitlement of all pupils to the CEG process:

Careers education and guidance have a central role in preparing young people for adult life . . . Good careers education and guidance are essential to pupils' and students' understanding and making choices which are right for them as they go through school and beyond.

HMI produced the widely acclaimed *Curriculum Matters 10: Careers Education and Guidance 5–16* (DES, 1988) which provided primary and secondary schools with a precise rationale for providing CEG programmes and, equally importantly, it gave clear guidance on implementing such programmes. HMI also took the opportunity to point out that the National Curriculum *per se* would not provide the careers education and guidance needed by pupils, when stating:

Careers education and guidance is needed by all pupils . . . and the ground-work for it needs to be carried out from the primary years onwards. To some extent it can be catered for within the attainment targets and programmes of study of the foundation subjects, but the specific careers education and guidance needed by pupils calls for a taught element that, if not in the Foundation curriculum, will need to be provided in the time not taken up by Foundation subjects.

The DES seemed to acknowledge HMI's concern when they issued *From Policy to Practice* (1989) in which it was declared:

The foundation subjects are certainly not the complete curriculum . . . More will be needed to secure the kind of curriculum required by Section 1 of the ERA. The whole curriculum for all pupils will certainly need to include at appropriate (and in some cases all) stages:

- careers education and guidance
- health education
- other aspects of personal and social education; and coverage across the curriculum of gender and multi-cultural issues.

In 1990, NCC published the long-awaited *Curriculum Guidance 6: Careers Education and Guidance* with detailed information on how the NCC saw the work of careers and guidance coordinators within the National Curriculum. It maintained the CEG momentum by stipulating:

Careers education and guidance is an integral part of the preparation of pupils for the opportunities, responsibilities and experiences of adult life . . . Much careers education and guidance can be provided within the subjects of the National Curriculum and other elements of the whole curriculum, and will be the responsibility of all teachers.

The Government's white paper *Education and Training for the 21st Century* (HMSO, 1991) has further highlighted the importance of CEG provision with proposals for better careers advice and closer collaboration between Training and Enterprise Councils (TECs) and LEAs regarding CEG provision. The extension of Training Credits and Compacts are also detailed in the white paper and as both initiatives are closely linked with CEG provision in schools, it lends yet more weight to the case for the development of effective CEG programmes.

Indeed, far from being the poor curriculum relation, CEG is rapidly becoming the natural focal point for many current educational initiatives; for example, individual action plans and the National Record of Achievement will quite obviously fit easily within a well organised and effective CEG programme.

The need for schools to provide a careers education and guidance programme for all pupils is, therefore, indisputable, but for this programme to be effective, careful planning and preparation must take place. These issues will now be dealt with in some detail.

WHAT SHOULD THE CEG PROGRAMME WE OFFER OUR PUPILS CONTAIN?

There are four important publications relating to CEG which will help to effectively address this question.

1. Working Together for a Better Future (DES/DoE, 1987)

This short, readable, booklet was produced by the Department of Education and Science and the Department of Employment and provides a concise and easy-to-follow framework for developing a CEG policy. Its clear message is that in order to deliver effective CEG the various partners involved in the process need to have a clear understanding of their roles and responsibilities. It is emphasised that schools must not work in isolation when devising and delivering their CEG programmes and that significant contributions can be made by parents, LEA Advisers, the Careers Service, governors, councillors, trade unionists and employers. The booklet offers guidance regarding:

- who to turn to for help when planning CEG programmes;
- what contributions should be made by primary, secondary and special schools and colleges;
- the role of the Careers Service;
- the role of parents; and
- monitoring quality of provision.

2. LEA Careers Education and Guidance Policy

All LEAs were asked to respond to the DES/DoE (1987) document by providing details of their own CEG policies. In theory, therefore, such a policy document should exist in every LEA and if there isn't a copy in your institution then contact the Advisory Service or the Careers Service. It is important to enlist the support of the LEA via the Advisory or Careers Service because, as stated above, there are a number of partners to involve when planning and implementing your own CEG programme.

3. Curriculum Matters 10: Careers Education and Guidance 5–16 (DES, 1988)

This document is widely acknowledged as one of the most authoritative and useful documents in supporting schools to develop programmes for the teaching and learning of CEG. It focuses on the aims and objectives of careers education and guidance in *secondary schools* and the relationship of such work to the experiences provided in *primary schools*. It discusses the implications of these aims and objectives for the choice of content, for teaching approaches, for curricular organisation and for the assessment of pupils' progress.

HMI clearly had one eye on the National Curriculum and its likely implications for timetable congestion but they are unequivocal in their pronouncement that the groundwork for CEG must be carried out in the primary years and that during the secondary years time will be needed for a taught element of CEG that cannot be provided in the foundation curriculum.

The document is essential reading when planning a CEG programme, covering as it does:

- the nature and scope of CEG;
- aims and equal opportunities;
- objectives – knowledge, skills, attitudes and personal qualities;
- provision 5–16, including:
    the careers education programme
    the Careers Service
    work experience and industry links
    contribution of subjects
    guidance
    accommodation and services;
- assessment and
- criteria for CEG.

4. Curriculum Guidance 6: Careers Education and Guidance (NCC, 1990)

This, the most recent document on the essential reading list, aims to help institutions deliver CEG as a cross-curricular theme; it refers to CEG as an 'integral part' of a pupil's curriculum. Section 2 of the report summarises the general aims of CEG, reflecting the aims expressed in *Curriculum Matters 10* (DES, 1988), as well as re-stating that:

Careers education and guidance should help pupils to:
- know themselves better
- be aware of education, training and career opportunities
- make choices about their own continuing education and training and about career paths
- manage transitions to new roles and situations.

NCC further suggests five strands in pupils' development which should permeate the curriculum at each of the key stages and describes them as:

Self	–	knowledge of self; qualities, attitudes, values, abilities, strengths, limitations, potential and needs.
Roles	–	position and expectations in relation to family, community and employment.
Work	–	application of productive effort; including paid employment and unpaid work in the community and at home.
Career	–	sequence of roles undertaken through working life and the personal success, rewards and enjoyment it brings.
Transition	–	development of qualities and skills which enable pupils to adjust to and cope with change e.g. self-reliance, adaptability, flexibility, decision-making, problem solving.

Section 3 emphasises that a school's policy on CEG should be developed in the context of its policy on cross-curricular themes which itself should be part of the school's whole curriculum policy.

Section 4 identifies and explains the five key components of a careers education and guidance programme:

- careers education;
- access to information;
- experience of work;
- access to individual guidance; and
- recording achievement and planning for the future.

Section 5 outlines a number of ways of organising CEG in the curriculum and there are specific suggestions regarding the contributions which can be made to CEG by the primary as well as the secondary curriculum.

Section 6 discusses the contribution of the Careers Service and other partners in the planning and delivery of CEG. It reiterates the advice given in *Working Together for a Better Future* (DES/DoE, 1987) in terms of involving the various contributors and indeed it is more up-to-date in that it refers to the support available from such new initiatives as Compact and TECs.

The final section deserves careful scrutiny as it provides guidance on 'Information in Key Stages 1–4' and provides a useful framework on which to base a CEG programme. NCC attempt to map the CEG curriculum and to indicate relevant attainment targets in National Curriculum subjects at each of the four key stages. Of course, not all final reports were available when NCC conducted this exercise but schools could usefully up-date this analysis as statutory and non-statutory guidance are published in other subjects. It must also be remembered that subjects other than the foundation subjects can play a significant role in developing the NCC's five strands of pupils' development.

Curriculum Guidance 6 is clearly the most relevant document in terms of addressing CEG as a cross-curricular theme within the National Curriculum but the three other documents listed above are invaluable when planning the whole school policy.

WHERE DO WE BEGIN WHEN PLANNING AND IMPLEMENTING CEG ACROSS THE CURRICULUM?

A step-by-step guide to the planning and implementation of CEG across the curriculum is provided below, but before considering it a few words of warning are offered. First, it should be recognised that the planning process itself will take a considerable amount of time and indeed a twelve-month preparatory period prior to full implementation is not an unreasonable period of time given the ever-increasing pressures schools are experiencing at the moment. Far better to ensure that planning and review is thorough and involves all the partners than to rush into the implementation stage with a 'go for it' mentality.

A second point to bear in mind is that primary and secondary schools will find that they may already be delivering significant aspects of the CEG programme albeit in a disparate and uncoordinated fashion. In other words, pupils will not be losing out because the school is taking its time in the planning, review and implementation process. On the contrary, if the planning is thorough, the implementation will be far more effective and beneficial to the pupils, relatively quickly.

The implementation phase itself will need to be viewed as at least a two- to three-year process – don't expect to implement a whole-school curriculum CEG policy during the first year. The review meetings suggested as part of the on-going monitoring and evaluation strategy will no doubt lead to amendments in both practice and future targets.

A final point to bear in mind is that there is a great deal of available support to help with planning and implementing a CEG programme. Advisers, careers officers, parents, governors, employers and trade unionists should prove very useful.

Planning and implementing CEG across the curriculum

1. Decide school policy for CEG
- Agree what is meant by CEG;
- review literature relating to CEG;
- involve Careers Service, all staff, parents, governors and the wider community;
- produce and distribute school policy statement for CEG which reflects whole curriculum policy;
- identify CEG coordinator and support team;
- ensure senior management and governor support; and
- consider INSET support.

2. Discuss delivery models
- Discuss models for delivering CEG;
- identify possible sources of support for delivering CEG, eg, Careers Service, Compact, TEC;
- produce paper on delivery model(s) for senior management; and
- consider INSET support.

3. Review current CEG provision
- Review current curriculum to identify which aspects of CEG are already provided for in terms of the aims and objectives stated in the school CEG policy statement;
- review Records of Achievement, assessment, recording and reporting processes and use of individual action plans in order to establish links with the careers guidance process;
- review current links with the Careers Service;
- identify gaps in current provision of CEG;
- review progression issue relating to CEG both within the school itself and between primary and secondary or secondary and further education;
- enlist the support of the LEA Advisory Service and the Careers Service to undertake the review;

- analyse resources for CEG across the curriculum;
- produce a paper on completion of the review for senior management and governors; and
- consider INSET support.

4. *Agree action plan for implementing CEG across the curriculum*
   - Needs to be planned within the context of the school development plan;
   - identify contribution by subjects, curriculum areas, or year groups;
   - ensure that the issue of progression is addressed;
   - assign staff responsibilities for delivery and coordination;
   - identify resource implications;
   - set long-term and short-term targets;
   - enlist the support of the Careers Service in implementing the programme – draw up a school/Careers Service 'agreement' plan with annual targets;
   - produce a paper for senior management/governors and all staff which outlines the implementation process and action targets; and
   - provide INSET support for staff involved in the implementation.

5. *Plan monitoring and evaluation strategy*
   - Specify criteria by which success in attaining action targets can be assessed;
   - identify staff responsible for monitoring and evaluating the CEG programme and agree on a schedule of review meetings;
   - be prepared to amend action targets in the light of review meetings;
   - enlist the support of the Advisory Service and the Careers Service in the monitoring and evaluation strategy;
   - produce a paper for senior management/governors on 'Monitoring and Evaluation of CEG' in the school; and
   - consider INSET support.

CASE STUDIES

The two case studies attempt to show how a primary school and an 11–18 comprehensive school are endeavouring to implement careers education and guidance across the curriculum. They are not 'blueprints' for action and indeed, as was stated earlier, the implementation phase is an on-going process which will be subject to regular review and amendment. They do, however, represent concerted efforts by schools in the primary and secondary sectors to address the management of CEG.

This two-form entry primary school is situated in one of the shire counties. One of the teachers was appointed coordinator for cross-curricular themes and for a variety of good reasons the school chose to focus upon CEG.

1. Discussing issues – drafting policy

An INSET day for the whole staff was organised by the coordinator, with input from the Careers Service. They reviewed NCC6, HMI CM10 and the LEA's careers education and guidance policy statement. A school policy statement was drafted and the staff decided to use Section 7 of NCC6 as a framework for implementing the school's CEG objectives across Key Stages 1 and 2.

One of the school governors took part in the INSET day and agreed to join a working group charged with planning and implementing CEG across the curriculum.

2. Agreeing policy – planning ahead

A draft policy statement was presented by the head to a governor's meeting. The school coordinator and careers officer were invited to the meeting to give a short presentation. Slight changes were made and a CEG policy statement was agreed (see Figure 4.1).

The head, coordinator, careers officers and governor met to plan a series of twilight meetings in order to carry out the next stage.

3. Reviewing the current situation

Working with their year-group colleagues, teachers were asked to check their schemes of work against the school CEG policy statement objectives. Staff were asked to give contextual examples of where they felt the objectives were being addressed, however generally. The coordinator, careers officer and governor spent some time with each of the year teachers working through this exercise.

The head and the deputy reviewed the school's recording and reporting procedures and the extent to which pupils were involved in the process. The careers officer agreed to help the coordinator compile a list of resources, including posters, which could be used to support the CEG programme throughout the school.

A further INSET day was agreed in order to discuss the review phase and to agree the implementation strategy. The coordinator collected and collated outcomes of the reviews as and when they were completed.

4. Taking stock and planning the implementation phase

The INSET day for all staff was planned by the coordinator, careers officer and governor in consultation with the head. Year-group teachers reported back on their review and the head provided feedback on the analysis of the recording and reporting procedures used and the extent to which pupils were involved. The greater part of the day was then spent discussing specific proposals put forward by the head, coordinator and careers officer relating to:

- A progression scheme based upon NCC6 chapter 7, indicating objectives covered by all classes at each key stage (see Figures 4.2 and 4.3 for the outcomes of this work).
- Introduction of portfolios into which pupils would place relevant pieces of work from their topics which related to the CEG objectives. These portfolios would progress with the pupils through the school.
- The head put forward proposals relating to the pupil's review of progress with the class teacher – to take place at the end of each term using the portfolio as the basis for the review.
- The head also suggested changes to the report format, to take

Figure 4.1 *A CEG policy statement – primary school*

NEWTON PRIMARY SCHOOL

CAREERS EDUCATION AND GUIDANCE POLICY STATEMENT

Aim

Within the framework of our whole-school curriculum we shall seek to provide careers education and guidance which will help our pupils to:

- develop the skills, attitudes and abilities necessary for taking an effective part in adult society.

Objectives

In support of this over-riding aim we shall endeavour to provide opportunities for the pupils to:

- develop knowledge and understanding of themselves and others as individuals – their strengths, qualities, attitudes, potential needs, values and limitations;
- develop knowledge and understanding of the world in which they live and the employment, career opportunities and voluntary opportunities available;
- develop the skills and qualities to make informed choices in relation to future education, training, occupation and career opportunities;
- manage the transitions within school and between schools and also subsequently from education into adult life.

Figure 4.2 *A progression scheme for Key Stage 1*

AIM	ACTIVITY	YEAR	TOPIC/CONTENT
1A	To begin to form an impression of self	Reception	Myself: Growth and change
		1	Home
		2	People who help us
1B	To identify and describe work which adults do in the pupils' immediate environment	Reception	People who help us
		1	Night and Day
		2	People who help us
1C	To explore adult work roles	Reception	People who help us
		1	Home
		2	Occupations
1D	To examine ways in which work in the locality has changed	Reception	People who help us
		1	Our town
		2	Farming; Transport
1E	To gain experience of a work place outside the school	Reception	
		1	Our town
		2	Transport

account of DES guidelines on reporting to parents and to include an element of self-assessment for years 4, 5 and 6. (An example of a self-assessment sheet subsequently agreed by staff is shown in Figure 4.4.)

- The head had to make contact with three local secondary schools to discuss transfer of information and summative Record of Achievement documents.
- The careers officer presented examples of resources which could be used to help promote pupil awareness of occupations, careers, roles, etc. Class teachers would decide how and when to use these resources. The head agreed to target £200 of capitation allowance to the purchase of materials.
- The implementation phase would commence in September 1991 with all year groups attempting to integrate their objectives into their topics.
- The coordinator agreed to pull all the above developments

Figure 4.3   *A progression scheme for Key Stage 2*

AIM	ACTIVITY	YEAR	TOPIC/CONTENT
2A	To review personal experiences as a basis for setting new targets	3	Termly reviews
		4	Termly reviews
		5	Termly reviews
		6	Termly reviews
2B	To recognise and constructively to discriminate against certain social groups	3	Industries and occupations
		4	Farming
		5	The universe
		6	Welfare state
2C	To understand how work involves a variety of related tasks, undertaken by people with different roles	3	Industries and occupations
		4	Clothes, farming
		5	Local study
		6	Occupations and industries
2D	To identify ways in which different types of work are like and unlike each other	3	Industries and occupations
		4	Farming
		5	Local study
		6	Occupations and industries
2E	To contrast work in different cultures and at different times	3	Celtic Britain and neolithic occupations
		4	Saxons, Normans, Vikings
		5	Tudors, Stuarts, Victorians
		6	Early 20th Century; World Wars
2F	To explore and compare how adults feel about their work	3	Industries and occupations
		4	Farming
		5	Local study
		6	Occupations and industries
2G	To anticipate and plan for moving to secondary school	3	
		4	
		5	Secondary school visit
		6	Secondary school visit
			Termly review

together into a paper for governors and into a short article for the school magazine and school prospectus.

5. *Monitoring and evaluation*
Targets for the first year of implementation were agreed (see Figure 4.5). One twilight meeting per term during 1991–2 was to be devoted to reviewing the implementation of CEG.

Figure 4.4 *A self-assessment sheet*

Name: _____ Class: _____

1. How do you think you have worked in these topics?

   a)

   b)

   c)

   d)

2. Which topic have you enjoyed and why?

3. What have been your main problems with your work this year?

4. What are you going to try and change about your work next year?

5. How have you behaved this year in school?

6. Which school clubs or teams have you taken part in?

7. Have you been given any responsibilities in school?

8. What hobbies and activities do you take part in out of school?

Pupil Signature: _____

CASE STUDY 2

IMPLEMENTING CAREERS EDUCATION AND GUIDANCE ACROSS THE CURRICULUM IN A SECONDARY SCHOOL

This six-form entry 11–16 school is situated in an inner-city area. The recently appointed careers teacher has been given the responsibility of coordinating CEG across the curriculum for all year groups 7–11.

1. Decide school policy for CEG

A two-day INSET course organised by the Advisory Service and Careers Service was attended by the head and the careers coordinator. One of the activities included drafting a school policy for CEG.

A half-day INSET for all staff was then organised by coordinator and careers officer focusing upon HMI CM10 and NCC6 with reference to *Working Together for a Better Future* (DES/DoE, 1987). The draft policy

Figure 4.5 *Targets for the first year of implementation*

ACTION TARGETS

In order to develop a programme of careers education and guidance across the curriculum the following action targets have been agreed for year 1 of the implementation process:

- ensure all staff are aware of the school's objectives relating to careers education and guidance across the curriculum;
- identify topics/context for developing aims outlined in NCC6 chapter 7 as they relate to each year group in Key Stages 1 and 2;
- ensure all pupils have a portfolio into which they can place examples of their work which relate to the careers education and guidance aims;
- initiate termly reviews of progress with all pupils using the portfolio as the basis for discussion;
- introduce pupil self-assessment sheets with years 4, 5 and 6;
- work with secondary schools on developing a summative Record of Achievement;
- work with the careers officer on the provision of relevant resources and materials to support the careers education and guidance programme;
- provide INSET for all staff to help with the implementation, monitoring and evaluation of the careers education and guidance programme.

---

prepared by the head and coordinator was presented for discussion and amendments. A final draft was subsequently prepared by the coordinator for presentation at a governor's meeting (see Figure 4.6).

2. Discuss delivery models

The head convened a CEG group comprising the deputy head, CEG coordinator, RoA coordinator, head of PSE, TVEI(E) coordinator, heads of years 7 and 10, and heads of communications and science faculties. The group was charged with producing a delivery model for introducing CEG across the curriculum in the next academic year. The school careers officer was also invited to join the group. The group decided to audit existing practice before deciding on a delivery model.

3. Reviewing the current situation

An audit form was produced based upon the aims of NCC6 chapter 7. This was distributed to all members of staff and time allocated to two specific directed-time meetings for completion of the form in faculties and departments (see Figures 4.7 and 4.8). A review of profiling/RoA processes initiated through TVEI(E) was also undertaken.

The support of the Compact team and Careers Officer was enlisted

Figure 4.6   *A CEG policy statement – secondary school*

CAREERS EDUCATION AND GUIDANCE POLICY STATEMENT

Aim

Within the context of providing a balanced and broadly based curriculum which prepares our pupils for the opportunities, responsibilities and experience of adult life, we shall aim to ensure that all pupils experience a careers education and guidance programme which will enable them to:

- know themselves better;
- be aware of education, training and career opportunities;
- make choices about their own continuing education and training and about career paths;
- manage transitions to new roles and situations.

Objectives

- a planned programme of activities to develop careers education and guidance across the curriculum;
- to provide a programme of careers education and guidance which furthers pupils' awareness of their own potential academic ability:
- to increase pupils' understanding of individuality, differing characteristics, interests and abilities, and the way these affect career choice;
- to broaden pupils' knowledge of the educational and vocational opportunities available locally, nationally and in Europe;
- to increase pupils' critical awareness of careers information;
- to help pupils develop the skills and knowledge relevant to coping with transitions;
- to raise pupils' awareness of the gender and racial issues inherent in our society;
- to ensure that our pupils fulfil their potential irrespective of their race, gender, religion or disability.

---

to pursue the idea of introducing individual action plans with year 10 students. The school would become a 'pilot' Compact school next academic year.

The careers coordinator and careers officer reviewed the current CEG programme, the careers officer's access to pupil information, and the availability of careers interviews and statistics on post-16 destinations and careers resources. The TVEI(E) coordinator reviewed the work experience programme and the preparation and debriefing of pupils.

4. Planning and implementing the CEG programme
The CEG group collated and discussed all the information gathered

Figure 4.7 *An audit form for Key Stage 3*

		SUBJECT_____		
KEY STAGE 3				
AIM	ACTIVITY	YEAR GROUP	T	CONTEXT
3A	To strengthen knowledge of self	7		
		8		
		9		
3B	To participate in decision making that requires their own & other people's point of view to be taken into account	7		
		8		
		9		
3C	To explore the careers/experience of work of individuals admired by pupils	7		
		8		
		9		
3D	To identify local employment opportunities	7		
		8		
		9		
3E	To consider controversial issues related to work	7		
		8		
		9		
3F	To compare how people earn their living locally and nationally	7		
		8		
		9		
3G	To prepare for curriculum choices in KS4 taking account of implications for future career opportunities	7		
		8		
		9		

Note: T = Treatment ie. thoroughness of dealing with the particular aim in your subject area.
     1 = thorough treatment where appropriate, please refer
     2 = moderate treatment to ATs.
     3 = slight treatment

from the review and produced a progression plan indicating the contribution by curriculum areas to the CEG programme across Key Stages 3 and 4 (see Figures 4.9 and 4.10).

A one day INSET day was planned with input from:

- head – CEG and the school aims;
- careers coordinator – a model for CEG across the curriculum;

Figure 4.8 *An audit form for Key Stage 4*

	SUBJECT_____			
KEY STAGE 4				
AIM	ACTIVITY	YEAR GROUP	T	CONTEXT
4A	To strengthen understanding of the qualities required for teamwork	10		
		11		
4B	To prepare for situations in adult working life where negotiation and assertiveness may be required	10		
		11		
4C	To prepare for choices of education, training or employment post-16	10		
		11		
4D	To examine the inter-action between domestic and other work roles in adult life	10		
		11		
4E	To make personal contact with people in their work roles in the community and develop further an understanding of relationships at work	10		
		11		
4F	To explore the inter-national perspectives of work and career opportunities	10		
		11		
4G	To identify and examine sources of information about future career opportunities	10		
		11		
4H	To explore future work opportunities	10		
		11		
4I	To prepare for the tasks involved in obtaining further education training or employment	10		
		11		

- careers officer – the role of the careers officer in the school plan for CEG;
- TVEI(E) coordinator – work experience – aims, preparation and debriefing;
- RoA coordinator – individual action plans.

Each input was followed by discussion and feedback.

Figure 4.9   *A progression scheme for Key Stage 3*

AIM	ACTIVITY	YEAR	PSE	ROA	EN	MA	SC	TE	HI	GE
3A	To strengthen knowledge of self	7	*	*	*	O	*	O	O	O
		8	*	*	*	O	O	O	O	O
		9	*	*	*	O	O	O	O	O
3B	To participate in decision making that requires other people's point of view to be taken into account	7	*	*	*	*	*	*	*	*
		8	*	*	*	*	*	*	*	*
		9	*	*	*	*	*	*	*	*
3C	To explore the careers/experience of work of individuals admired by pupils	7	*		O				*	
		8	*		O		O			
		9	*		O			O		
3D	To identify local employment opportunities	7	*							
		8	*							*
		9	*		O		O			*
3E	To consider controversial issues related to work	7	*		O		*			
		8	*		*	O	O	*		O
		9	*		*		*	*		*
3F	To compare how people earn their living locally	7	O							
		8	O				O			O
		9	*				O			O
3G	To prepare for curriculum choices in KS4 taking account of implications for for future career opportunities	7	O	*						
		8	O	*						
		9	*	*	*	*	*	*	*	*

Note: * denotes thorough treatment   O denotes moderate treatment.

The CEG group met to review the INSET day and to amend the progression plan and implementation process. A paper was produced for presentation to the governors by the head, outlining plans agreed for delivering the CEG programme commencing in the next academic year. Resource implications were outlined together with INSET plans to support the implementation phase.

5. *Monitoring and evaluation*
Performance indicators for the first year of implementation were agreed by CEG group and the head. Four review meetings during the

Figure 4.10 *A progression scheme for Key Stage 4*

AIM	ACTIVITY	YEAR	WORK EXP.	CAREERS	PSE	ROA	EN	MA	SC	TE	HI	GE
4A	To strengthen understanding of the qualities required for teamwork	10	•	•	•	•	•	o	o	•	o	•
		11		•	•	•	o	o	o	o	o	o
4B	To prepare for situations in adult working life where negotiation and assertiveness may be required	10	•	•	•	•	•		o	o	•	•
		11		•	•	•	o			o	o	o
4C	To prepare for choices of education, training or employment post-16	10	•	•	•	•	•					
		11		•	•	•	•					
4D	To examine the interaction between domestic and other work roles in adult life	10	•	•	•		o					
		11		•	•		o					
4E	To make personal contact with people in work roles in the community and develop further an understanding of relationships at work	10	•	•	•	•		o	o			•
		11		•	•		o		o		o	o
4F	To explore the international perspectives of work and career opportunities	10	•	•	•						▯	•
		11		•	•							o
4G	To identify and examine sources of information about future career opportunities	10	•	•	•	•	o	o	o	o	o	o
		11		•	•	•	o	o	o	o	o	o
4H	To explore future work opportunities	10	•	•	•	•	o	o	o	o	o	o
		11		•	•	•	o	o	o	o	o	o
4I	To prepare for the tasks involved in obtaining further education training or employment	10	•	•	•	•	•	•	o	o	o	o
		11		•	•	•	o	o	o	o	o	o

Note: * denotes thorough treatment
O denotes moderate treatment

next academic year were scheduled for the CEG group specifically to review progress and amend plans where necessary.

A CEG 'contract' was drawn up between the school and the careers officer indicating the targets and responsibilities of both parties in terms of implementing and monitoring the school's CEG plans. The 'contract' would be reviewed annually with the head and the CEG group.

SUMMARY

CEG across the curriculum plays a vital role in ensuring that pupils are adequately prepared for the opportunities, responsibilities and experiences of adult life.

In planning and implementing their programmes schools need to involve governors, parents, employers, LEA advisers, careers officers and the wider community and the starting point for developing a cross-curricular approach to CEG has to be a school policy statement which is owned and understood by all staff.

A staged development phase for introducing CEG across the curriculum at all Key Stages should be based upon:

- deciding school policy for CEG;
- discussing delivery models;
- reviewing current CEG provision;
- agreeing an action plan for implementing CEG across the curriculum; and
- planning a monitoring and evaluation strategy.

REFERENCES

DES (1985) *Better Schools*, Cmnd 9469, London: HMSO.
DES (1988) *Curriculum Matters 10: Careers Education and Guidance, 5–16*, London: HMSO.
DES (1989) *National Curriculum: From Policy to Practice*, London: HMSO.
DES/DoE (1987) *Working Together for a Better Future*, London: HMSO.
NCC (1990) *Curriculum Guidance 6: Careers Education and Guidance*, York: NCC.
HMSO (1991) *Education and Training for the 21st Century*, London: HMSO.

Chapter 5

Economic and Industrial Understanding – The Mini-market

Alan Blyth

ORIGINS

One aspect of educational policy that developed with increasing clarity during the 1980s was the need for schools to prepare children and young people for the world of work.

Several different interests were involved in this development: employers anxious to recruit good-quality workers; trade unionists concerned with the image and status of 'real jobs'; politicians, especially those committed to the promulgation of an enterprise culture; and educationists convinced of the importance of relevance in the curriculum. For the time being, at least, a coalition between these interests, backed by considerable public and private funding, ensured that attention would be paid to issues concerned with the world of work, first in secondary and further education but, as confidence was developed, in primary education too. What had begun here and there as individual innovations grew until it became an 'entitlement', part of the curriculum to which every child could lay claim. The terms 'industry', 'enterprise', 'economics', 'awareness' and the more comprehensive 'understanding' (awareness plus knowledge) all came to be used by various participants in the accelerating drama.

When the idea of cross-curricular themes emerged, as part of the new curricular design symbolised by the Education Reform Act of 1988, it was likely, in view of a decade or so of innovations, that something in this general area would qualify as one of these themes, and so it did. Three of the current terms were eventually chosen to designate it: economic and industrial understanding (EIU). Soon this was recognised as a significant part of what Dufour (1990) has called the 'new social curriculum' across the age range 5–16, and before and after those ages too. However, some approaches to EIU had become characteristic of the age groups 5–13, and it is those that will be discussed in the present chapter, under a familiar title adapted for this purpose: 'the mini-market'.

THE NATURE OF THE MINI-MARKET

It is necessary to take a closer look at what the mini-market is about. The clearest semi-official statement is from the National Curriculum Council (NCC, 1990) itself: *Curriculum Guidance 4: Education for Economic and Industrial Understanding,* now usually known as CG4. It is a helpful and substantial document, mercifully free from the bland generalised agenda that masquerades as content in too many guides designed for teachers. It draws on the considerable repertoire of innovations carried out with children aged 5–16 and includes substantial treatment of the first three Key Stages. In structure it resembles the *Curriculum Matters* series issued by H.M. Inspectorate,[1] in that it spells out what is to be expected of children at different key stages, rather than positing attainment targets or programmes of study which can in fact only be legally assigned to core and foundation subjects. Unlike the more rigidly controlled aspects of the National Curriculum, it also makes positive if uncontroversial recommendations about attitudes. The result is that the mini-market is given greater, not less, flexibility and is spared some of the contradictions that those subjects are likely to meet.[2] Accompanying the basic layout of these expectations of educational outcomes at different ages are good examples of how to organise activities (not programmes of study) calculated to promote those outcomes, with actual case studies of children at work.

Throughout CG4, however, one issue is evaded, perhaps deliberately. As in surprisingly much of the literature in this field, even in the title of Ross's splendid survey (Ross, 1990), it is assumed that economic-and-industrial-understanding is a unitary concept. Now and again, the adjective 'economic' is used separately, but the next moment the authorship of CG4 remembers to use the full title. Yet these two adjectives are not interchangeable. As Pearce (1988) indicates, they are not even in the same category. Economics is a discipline in its own right, with a related school subject at the secondary (and further) level involving a way of thinking that has a definite place in the primary years too, although it does not have the status of a core or foundation subject. It belongs to the family of disciplines that includes history, geography and the other social sciences. Incidentally, EIU is the only one of the cross-curricular themes to be based on a distinctive discipline. Industry, meanwhile, is a field of human activity to which economics is obviously important, but so are science, technology, art and other studies. In the usage found in CG4 and other literature, 'industry' is sometimes used just to refer to manufacturing industry, and sometimes to cover other forms of economic activity that depend on different kinds of expertise: agriculture, commerce and the tertiary and quaternary sectors (Ross, 1988; Smith, 1988). Even where there is quite close correspondence between economic principles and industrial practice, the first two elements in EIU remain conceptually distinct, in much the same way as mechanics and motoring are distinct.

That distinction becomes more evident when considered in relation to the third part of the title, 'understanding', which is in turn coupled with the other two without explanation. Understanding in economics means grasping basic principles and their elaboration; recognising the validity of generalisations about an aspect of human behaviour; subsequently making measurements and predictions about that behaviour; and being aware of the practical relevance of this way of thinking. Understanding industry means getting to know how a particular organisation works, what technology it uses, what kind of people work there, why they work there, how they relate to each other, what that particular enterprise 'feels' and sounds and smells like and how it fits into local, regional and national life: education *about* and *through* industry (Blyth, 1984b). This second kind of understanding is more factual, more concrete and more cross-curricular than is economic understanding, reaching out also into the field of attitudes as well as of cognition. Both are important, but they are not identical.

If economics represents the more basic set of concepts to be fostered through EIU, industry (in the broadest sense) provides a splendid focus for cross-curricular learning. If economics provides the discipline for the mini-market, industry provides its programme. It is a topic, in fact a rich repertoire of topics, rather than a subject in itself, a consideration that will be developed further, below.

ARGUMENTS ABOUT THE MINI-MARKET

Hitherto, it has been assumed that some form of mini-market should figure in the curriculum 5–13, for reasons such as those outlined above. Those reasons are weighty, but they can be reinforced by another and more powerful one. For if the mini-market is an entitlement for every child, that is not because children are potential employees, or business people, or consumers, or even citizens, but because they are already sentient, developing human beings who are entitled to economic insights and knowledge of industry in the same way as to all the rest of a full curriculum. It is part of the total environment that they meet as they grow (Ford, 1989–90; Ross, 1988) and can contribute distinctively to the development of thinking skills (Craft, 1990) through the enabling intervention of teachers (Blyth, 1984a). Not everyone would agree that the mini-market should be in the primary curriculum at all. Various objections could be raised, and these must be given due consideration.

The composite concept of the mini-market – economic and industrial understanding – is difficult to justify

This difficulty can best be met by a frank awareness of the relationship between the two elements in the title and by skilful out-working from the children's own first-hand experiences. Such difficulties arise most readily when a teacher decides to teach what she defines as economics, rather than to encourage the development of economic ideas, for example through some topic in local economic and social activity.

There is too much disagreement about the purposes of including the mini-market in the curriculum

There is indeed diversity among the advocates of the mini-market, and teachers need to be aware of this. However, as Pearce (1989) has shown, it is possible to harmonise the claims of economic literacy, experiential learning and social needs, provided that the focus of attention for younger children is confined to examples that are readily comprehensible.

Children 5–13 are too young to understand the mini-market or to tolerate its ambiguities and its controversial nature

This is a Peter Pan view of the mini-market, which is not borne out by experience. Children do prove able to learn, progressively, what the mini-market is about. That, indeed, is the basis of the descriptions of children's capabilities at different age-levels in CG4. Moreover, children from 9 onwards *should* be introduced to the need to handle controversy and to tolerate ambiguities.

Children 5–13 already understand the mini-market only too well

The argument here, almost the exact opposite of the previous one, is that working-class children are necessarily streetwise about how to handle scarcity, and middle-class children understand the importance of money through their family experiences, so that the idea of incorporating the mini-market into the curriculum will be seen by both as superfluous and boring: nobody likes having their own experience laboriously explained to them. This is an understandable concern, but where the mini-market has operated, there has been too much richness of potential to allow for any onset of boredom or complacency. There is always so much more to learn.

The inclusion of the mini-market in the curriculum could be seen as legitimating the pursuit of purely material ends

In this view there is a fear that children are already too prone to accept materialistic values and that the mini-market will only aggravate their self-interest. This is an ethical argument, related to the plurality of purpose mentioned above. It can, and should, be answered through a morally responsible approach to the mini-market. It is both possible and necessary to convey the impression that humankind does not live by bread alone. At the same time it is both possible and necessary to include the understanding of economic behaviour within the scope of learning.

There simply isn't room for the mini-market in the already overloaded curriculum in Key Stages 2 and 3

The stock answer to this, as was implied earlier, is that it is a part of a cross-curricular theme, a 'bonus issue' that should permeate the core and foundation subjects without taking up any extra time. To be candid, that answer often fails to convince. It is preferable to assert that, somewhere in each of Key Stages 1, 2 and 3, room must be found for some, but only some, explicit treatment of the mini-market, while its importance is remembered also in other topics with a historical, geographical, technological or scientific emphasis. More than this would tilt the balance too far towards the mini-market, and might indeed induce the very element of boredom previously mentioned. To avoid this, and indeed most of the other objections, it is necessary to plan consistently through a whole-school policy for continuity and progression in learning and indeed in assessment.

PLANNING FOR THE MINI-MARKET

Thanks to the initiatives taken in the 1980s, there is no lack of advice to teachers and others intending to set up the mini-market in an infant school, a junior school, a middle school or the first years of a secondary school. In addition to CG4, there are four volumes entirely concerned with the mini-market: Hutchings and Wade (1991); Ross (1990); Smith

(1988) and Smith (1991) and other useful but less accessible compendia of case studies (Smith, 1987; Waite, 1987) which, like these four, have been used as sources when formulating the suggestions that follow. Other sources well worth consulting as supplementation to CG4 are chapters like the present one which refer to the mini-market in a wider context (for example Ross, 1989) and articles in *SCIP News*, the journal produced by the School Curriculum Industry Partnership (as it now is). Research is continuing in several centres which periodically publish findings.[3] There are also various 'packs' prepared by organisations active in this field.[4] Still other material concentrates on the place of the mini-market in the professional education of primary and secondary teachers.[5] So there is plenty of advice and assistance about what to do, much of it emanating from the same handful of enthusiastic pioneers. For that reason, the emphasis here will not be on this thesaurus of exciting work with children, but rather on the planning of such activities on a whole-school basis within the National Curriculum.

Such planning, with due regard to continuity and progression in learning, has not always been as evident as it should be.[6] Pragmatically, there is a case for including one such venture for all children in the infant years and another in the junior years, leading into a somewhat different treatment at 11–13. This requires from the schools' management a degree of whole-school agreement within each school, and of cooperative planning about policy between schools concerned with different key stages. Here, CG4 offers a very useful framework of suggestions, but it has to be clothed with detail in each individual case.

The work at *Key Stage 1* (5–7) offers the greatest flexibility and is thus, in one sense, the easiest to plan. On the other hand, economic concepts are likely to be at their most rudimentary, while the range of opportunities afforded by industry as a topic field is also necessarily limited, though it can be imaginatively explored by visits, simulations and very simple mini-enterprises: 'micro-enterprises'. The *Shops* topic outlined in CG4 indicates some of the possibilities that exist: here the real mini-market can be encountered. Other themes that have been pursued successfully include a local bakery, manufacturing industries, a farm and even a small port. This list could be extended to include other readily comprehensible economic units: a fishing boat, a local coach operator, a garden centre and so on, each with cross-curricular links that suggest themselves immediately. What the National Curriculum demands, however, is that these cross-curricular links, like the relevant economic concepts themselves, should be seen not as opportunities to drag as many core and foundation subjects in as possible, but rather to ensure that the natural links with a coherent range of those subjects should be pursued in such a way as to promote progression in the attainment targets in those subjects, and development in the other cross-curricular themes and dimensions, notably language. At the same time, the main emphasis should be on how the actors in this little drama plan to make their enterprise viable and how costs, opportunity costs, overheads, demand, marketing and division of labour figure in their activities; also the elements of conflict as well as of cooperation to which those activities give rise. Taking into account all these considerations, and the general guidance now being given in the various programmes of study, it seems preferable, though not essential, to concentrate work on the mini-market with six-year-olds rather than with five-year-olds: in year 2 rather than in year 1, while paying attention throughout Key Stage 1 to the development of economic ideas through activities in all aspects of the curriculum.

For each child, the learning accruing from such work must depend on the quantum of economic understanding with which he or she begins, and so although they all share a common experience and should be expected to acquire some common information, each will also have an individual starting-point where economic understanding is concerned, and a goal that emerges as the work proceeds. This must be taken into account in any assessment policy that is introduced.[7] Some will indeed proceed faster than others, and this will involve a decision in principle about how desirable it is that children should move at different paces. Personally, I believe that each should be taken as far and as fast as they can go, which is, as HMI have often claimed, farther and faster than some teachers have believed possible. But that is a matter for professional, even for ideological, choice.

Meanwhile, whatever purposes are embedded in the curriculum, the progress of Key Stage 1 children in understanding the mini-market can best be monitored if this is recognised as a legitimate and necessary part of what figures in their Records of Achievement. It is quite inadequate to leave it to figure as an offshoot of assessment in the core subjects.

At *Key Stage 2* (7-11) the scope of the mini-market widens substantially. Children have more capacity to grasp scientific principles, to engage at their own level in technological innovation, and to see enterprises in a wider historical and geographical perspective. They are also more able to understand, and to participate in, the kind of group behaviour that is involved in a simple social organisation. For this reason the mini-enterprise, an addition to the pedagogical repertoire initially devised for upper secondary and further education,[8] has now become established in junior (and in some infant) schools and there is already a quite impressive record of work built around that focus (HMI, 1990). Other significant approaches to the mini-market at this stage include more searching case studies and the introduction of issues concerned with local planning.

Here is an example of a mini-enterprise at this age level. A teacher and class exploring the potentialities of a material such as clay, and blessed with access to a kiln, might decide to try designing, making and marketing simple clay pots. It takes little imagination to see how such a project would relate to the attainment targets in technology and science and art and, more indirectly, to history, through the significance of ceramics as evidence of past civilisations, and in geography through the sources and movement of materials and the location of potteries. The essence of the mini-market itself would emerge through discussion of the economic issues considered at Key Stage 1 and wider ones too: market research, profit and loss, supply and demand, pricing, quality control and elementary accounting, all of which involve rudimentary management, decision-making, and even industrial relations shot through with oral English and the uses of mathematics.[9]

As a cross-curricular venture, such a mini-enterprise has obvious potential, as would other approaches to economic understanding. The problem, inevitably, is that of finding a niche for it in the curriculum when the demands of the core and foundation subjects themselves are so stringent. Probably the best age-level for this purpose is year 5, the nine-year-olds, with the focus in the time-equivalent allocated to technology but with some relevance to historical and geographical development and to three-dimensional work in art. The difficulty here is that pottery, for example, can only figure in the programme of study in history and geography if pot-making is and has been part of the local scene. Certain other economic activities such as farming, land transport and printing have more scope in the history programmes, while the first two blend quite well with some, though not all, home-region and environmental study in geography. So, regrettably, any mini-market topic has to be introduced into this National Curriculum with considerable dexterity.

Assessment of such work would follow on lines similar to those suggested for Key Stage 1. The emphasis would be on how far each child had developed, and what new concepts, ideas, skills and enthusiasms had appeared, rather than on the factual information gained through the topic itself. For this reason, once again, Records of Achievement, embodying the outcome of a probing assessment of children's capacity to grasp and adapt concepts, are far more likely to do justice to the mini-market than would ever be achieved by recall testing. Nevertheless, quick-fire tests of this recall kind do in fact have a minor role to play as a means of reinforcing learning of the material that has constituted a class's common experience.

In *Key Stage 3* (11–14), the limits of the mini-market begin to appear.

For now the subject structure of the curriculum becomes, legitimately, more marked. The practice of activities such as making, buying and selling gives way to a more systematic understanding of how an industry or a commercial organisation works, and of ideas such as capital investment and marginal cost. It is within geography, science, mathematics, technology or art that economic ideas can be most readily extended, as CG4 indicates, though there are ways in which economics impinges on every one of the core and foundation subjects – witness the importance of economic factors in history, the significance of English and other languages in trade but also in the interpretation of experience, and such matters as the organisation of the pop and leisure industries. Moreover, there is still a place within each subject area for an investigative approach to learning, rather than for the mere acquisition of pre-arranged content.

In essence, Key Stage 3 constitutes a bridge between the fully cross-curricular approach pursued in primary schools and the more mature EIU of the upper-secondary and further education years when comprehension can be more readily extended to macro-economic considerations including the national economy itself. On the other hand, as that transition continues, EIU continues to overlap more clearly with other cross-curricular considerations, notably citizenship, environmental education, equality issues and the one that in a sense grows out of EIU itself, namely careers education and guidance. Elements such as work experience then begin to appear in the programme. Key Stage 3 leads into these developments by keeping the significance of EIU constantly in mind in the various programmes of study, and by encouraging the raising of economic and social questions, even (or especially) when they are awkward. Meanwhile, although some aspects of assessment now become more formal, economic considerations should continue to figure in teacher assessment and – though this is not within the schools' individual gift – in any standard tests that may be introduced.

These are just a few of the considerations that will need to be borne in mind when curricular discussions take place within schools, and in clusters of schools in an area catering for children from 5 to 13. It is within such general planning that the mini-market can take a constructive place. Some measure of agreement between teachers and schools will be necessary, even if that implies some teachers curbing their preference for doing just what they fancy, or what they have always done. That degree of corporate planning and responsibility can only be advantageous. Incidentally, it can itself be made to illustrate the virtues and the limitations of *laissez-faire* behaviour. This kind of planning can be made to work, and work successfully, within the National Curriculum that we have. It might work much better in another that we may hope to have, one less hastily constructed, more effectively based on how children grow and learn, one in which 'mini-market forces' are given their due place as part of developing experience, though not of ultimate purpose.

NOTES

1. This series was issued by HMI before the Education Reform Act was passed and has some, but not a decisive, influence on the curricular provision of that Act.
2. Contradictions will emerge between the various elements in the statutory curriculum: attainment targets versus programmes of study and both versus levels and statements of attainment. This will be a result of the inordinate haste and lack of professional advice with which the main structure of the new curriculum was set up.
3. Notably at the Primary Industry Centre at Edge Hill College of Higher Education, the Primary Schools and Industry Centre at the Polytechnic of North London and the Centre for Education and Industry at the University of Warwick; also some others listed in CG4, Appendix, pp. 52–4.
4. Particularly useful packs for primary schools are produced by some of the organisations in the Appendix mentioned in the preceding note, especially, the Durham University Business School and the Educating for Economic Awareness Project.
5. See the publications from the Enterprise Awareness in Teacher Education project: also the journal *Economic Awareness* published by the Economic Awareness Teacher Training Project (EcATT), and the general literature on economics education by writers such as Whitehead and Ryba.
6. But see examples of structuring cited in Ross (1990), especially chapters by King, Schug and Ryba.
7. For detailed suggestions about assessment in the mini-market see Blyth (1988) and Blyth and Wilkins in Smith (1991); also (on primary humanities generally) Blyth (1990).
8. For a general discussion of the mini-enterprise 'movement' see Turner and Crompton (1991).
9. CG4's treatment of Key Stage 2 is centred on the relationship of the mini-market to the development of mathematical understanding.

REFERENCES

Blyth, W A L (1984a) *Development, Experience and Curriculum in Primary Education*, Beckenham: Croom Helm.
Blyth, W A L (1984b) 'Industry education: case studies from the North West' in Jamieson, I (ed.) *We Make Kettles: Studying Industry in the Primary School.* London: Longmans.

Blyth, W A L (1988) 'Appraising and assessing young children's understanding of industry' in Smith, D (ed.) *Industry in the Primary School Curriculum: Principles and Practice*, London: Falmer Press.

Blyth, W A L (1990) *Making the Grade for Primary Humanities*, Buckingham: Open University Press.

Craft, A (1990) 'Economic and industrial understanding and skills for thinking', *SCIP News*, 28 (Winter) 40–2.

Dufour, B (ed.) (1990) *The New Social Curriculum: A Guide to Cross-curricular Issues*, Cambridge: Cambridge University Press.

Ford, K (1989–90) 'Economic and industrial understanding in the Early Years curriculum', *Educating for Economic Awareness Forum* (Winter), 5–8.

HMI (1990) *Mini-enterprise in Schools: Some Aspects of Current Practice*, London: HMSO.

Hutchings, M and Wade, W (1991) (*Developing Economic and Industrial Understanding in the Primary School*, London: Polytechnic of North London Press.

National Curriculum Council (1990) *Curriculum Guidance 4. Education for Economic and Industrial Understanding*, York: NCC.

Pearce, I (1988) 'Economic and industrial awareness: "kinship partners" in the schools–industry field?', *SCIP News* (Winter) 16–17.

Pearce, I (1989) 'Putting perspectives in perspective', *Educating for Economic Awareness Forum*, (Spring) 3–10.

Ross, A (1988) 'Children's understanding of the social and economic world' in Smith, D, *op. cit*, ch. 2.

Ross, A (1989) 'Industry and Humanities' in Campbell, J and Little, V (eds) *Humanities in the Primary School*, London: Falmer Press.

Ross, A (ed.) (1990) *Economic and Industrial Awareness in the Primary School*, London: Polytechnic of North London Press.

Ross, A and Smith, D (1985) *Schools and Industry 5–13: Looking at the World of Work: Questions Teachers Ask*, London: SCIP and School Curriculum Development Committee.

Smith, D (ed.) (1987) *Industry Education in the Primary School: Case Studies from the National Primary Schools Industry Competition 1986*, London: School Curriculum Development Committee and Department of Trade and Industry.

Smith, D (ed.) (1988) *Industry in the Primary School Curriculum: Principles and Practice*, London: Falmer Press.

Smith, D (ed.) (1991) *Towards Curriculum Progression: Developing Economic and Industrial Understanding in the Primary School*, Coventry: SCIP/MESP Publications.

Turner, D and Crompton, K (1991) *Enterprise Education in the National Curriculum: Agenda for the 90s*, London: Community Service Volunteers.

Waite, P E (ed.) (1988) *Primary Schools and Industry Year 1986: A Report of the Industry Year Workshops*, London: Industry Matters.

Chapter 6

Economic and Industrial Understanding 14–19 – Adding to the Clutter or Broadening the Experience?

Ray Derricott

This chapter concentrates on the impact of teaching for economic and industrial understanding (EIU) at Key Stage 4 of the National Curriculum and beyond into the post-16 years. This focus was chosen because it is in these years that much investment through, for example, the Technical and Vocational Educational Initiative (TVEI), Understanding British Industry (UBI), the Schools Curriculum Industry Project (SCIP), Project Trident and various schools and business partnerships, had already been directed before the advent of the 1988 Education Reform Act (ERA). In response to such initiatives many schools and colleges have therefore already set up curricular structures and organisations to incorporate what they would call EIU into their programmes. Does the writing of EIU into the requirements for the whole curriculum simply formalise the place of these initiatives and indicate that they should be part of the normal entitlement of all pupils or does it, particularly at Key Stage 4 of the National Curriculum and beyond, interfere with well-established arrangements and make the problem of curriculum management and implementation much more complex? A tentative judgement on this issue will be made at the end of the chapter.

Another reason for concentrating on the 14 to 19 age group is that the problems of continuity or discontinuity that may occur pre- and post-16 can be examined. The statutory entitlement to a National Curriculum ends at 16 and, at the moment, there is no such broad framework for a curriculum laid down for students who stay in full-time or part-time education beyond this age. However, as early as 1989, the Further Education Unit issued its document *Towards a Framework for Curriculum Entitlement* (FEU, 1989) in which all those involved in post-16 education and training were exhorted to produce mission statements which make clear their purposes and emphases. In particular, colleges were asked to take on the implications of the 'client-centred college' and the 'client-centred curriculum'. The latter gave emphasis to core skills and understanding. In 1990, the National Curriculum Council entered the debate about what might be the post-16 consequences of the 5–16 National Curriculum in response to a question set for them by the then Secretary of State. The NCC's (1990a) document *Core Skills 16–19* emphasised the need to consider continuity from 5 to 19 and 'the need to bridge and ideally eradicate the divide between academic and vocational qualifications' (p. 3).

The NCC also claimed in their consultations to have met a wide consensus that the whole of the post-16 curriculum should be permeated with common themes, guidance and core skills. The five cross-curricular themes of the National Curriculum were to be augmented by two additional ones, namely, scientific and technological understanding and aesthetic and creative understanding.

The core skills to be incorporated into the curriculum of 16- to 19-year-olds were mainly those that had been well trialled and developed under TVEI and personal and social development programmes in the compulsory school years. The core skills were communication, problem-solving, personal skills, numeracy, information technology and modern language competence. The last was probably added because the National Curriculum makes the study of a modern language compulsory at Key Stages 3 and 4. The NCC went on to advocate the development of attainment targets for each of the core skills and to suggest ways in which the core skills could be incorporated in A and AS syllabuses. Communication, problem-solving and personal skills *should* be part of all post-16 programmes and *embedded in* every A and AS syllabus. Numeracy, information technology and modern language competence were more difficult to incorporate in every A and AS course but the NCC advocated exploration of possibilities. All activities could be reported in a Record of Achievement.

The NCC document on core skills was intended to promote discussion and quite clearly was only advisory but it was progressive in advocating change in that most conservative of educational arenas, that of the A level. The report quotes the general criteria for GCSE subjects which lay down that syllabuses should have a relevance to life and should promote awareness of economic, political, social and environ-mental factors. Similar criteria, it suggests, should be incorporated in the guiding principles for A and AS examinations. More fundamentally, the NCC recommend the development of *competences* related to the themes and the core skills and in this way they seem to be taking a strong stance towards the vocationalisation of the curriculum for 14- to 19-year-olds. Indeed, the current piloting of a personal competences model with students aged 14 to 22 is a further indication of the increasing influence of industrial management and training upon mainstream education.[1]

The persuasive and mainly ideological battle about the country's need for a more vocational approach to education seems now to have been won by 'the establishment'. In the early days of TVEI it was possible to read the strong arguments against vocationalism from such writers as Holt (1982) and Jonathan (1985). A useful analysis of these arguments is to be found in Murray (1989) who explores the language of official documents written to persuade the educational community of the worth of vocationalism. Summarising the ideology with which the White Paper on Education and Training is imbued, Murray concludes:

> Engendering team spirit, the White Paper reminds us that we are partners, 'working together' with 'motivation initiative and enter-prise' for the collective good of the team and the nation (p. 32).

The rhetoric about the 'why' of vocational education is strong but all the above issues raise questions about how the ideas and principles are going to be incorporated at the policy and practice levels in schools and colleges. The problems posed for curriculum management in the face of multiple innovation are great as schools and colleges attempt to cope with the introduction of CPVE and of AS levels alongside the development in pre-vocational qualifications by BTEC, RSA and CGLI and the establishment since 1986 of the National Council for Vocational Qualifications (NCVQ) as well as, at Key Stage 4, a programme which satisfies the requirements of the National Curriculum. What is paradoxical about this situation is that the arguments supporting the need for a national curriculum framework were in part about getting rid of what Sir Keith Joseph called 'curriculum clutter'. Education for EIU seemed to fit naturally into a technical and vocational initiative which encouraged cross-curricular initiatives, but how is it able to fight for space within the statutory curriculum which has a heavy subject-centred bias and where cross-curricular themes and dimensions become everybody's and therefore nobody's concern? Successive secretaries of state have indicated that there is no problem. Kenneth Baker in his Manchester speech in September 1987, referring to cross-curricular themes, said:

> . . . describing the curriculum in terms of subjects will not exclude these desirable areas of learning and experience. . . . [These were]

expected to be taught through other subjects, giving added dimension to what is taught, as most of them are now in the most effective schools.

In the same speech Baker went on to refer to the significance of TVEI:

TVEI has many lessons for schools about the improvement of these [cross-curricular] aspects of the school curriculum. The TVEI projects are providing valuable experience in identifying the most effective ways in which the education of 14- to 18-year-olds can be made more relevant to the demands and opportunities of employment . . . schools will need to use TVEI to build on the framework offered by the national curriculum, and to forward its objectives (Baker, 1987).

Thus, to the politician, in persuasive mode, TVEI, the National Curriculum and EIU are not in conflict, the best schools are already coping and the rest will need to look to their laurels. Just over three years after the above speech, John MacGregor had become Secretary of State and appeared to be in much less of a hurry over the implementation of educational reform than his predecessor. In January 1990, in an address to the Society of Education Officers, he dealt specifically with problems at Key Stage 4. He saw these as overcrowding, fragmentation and certification and he rehearsed the, by then, well-known arguments about the pressure on teaching time from the National Curriculum requirements and the need to provide pupils with

room for options, and opportunities for pupils of varying aptitudes to obtain a sensible number of good GCSEs . . . [while at the same time avoiding the possibility] that the curriculum will become split into undesirably small blocks of work which will not motivate pupils (MacGregor, 1990).

His solution amounted to a recommendation to combine courses, to allow pupils who had already reached level 8 by Key Stage 4 to drop subjects and to advocate the vocational qualifications of BTEC, RSA and CGLI.

To attempt to assess what effects this rhetoric and general advice is having 'on the ground' the author interviewed the head of a comprehensive secondary school in the North-West of England and used evidence from a case-study of a business/education partnership project which was based on a smaller 11 to 16 comprehensive secondary school. Participants were assured of anonymity and therefore names of individuals, schools, companies and specific locations have not been used.

The head was asked the general question: 'How are you coping with the demands of the National Curriculum and the cross-curricular themes and dimensions, particularly EIU, especially at Key Stage 4?'

H: How long have we got? I find it difficult to focus on the question the way you have asked it . . . I'm not sure I can answer it . . . perhaps you've got the wrong person. You should see X . . . As you know I've not been here too long and so many of the structures we're operating are what I found here. Fortunately, he (the previous head) had made no decisions about the NC – I say fortunately because anything he did would probably have been wrong – I don't mean that critically but I know too many colleagues who have tried to second-guess the Secretary of State's intentions and they've put too much unnecessary pressure – stress – on to their staff . . . so I'm not trying to justify inactivity – not at all. . . . I can only concentrate on having people in place who can think things through, have some general guiding ideas – principles if you like – and work things out from there. . . . We're into TVE here – you could almost say we take it for granted. . . . I expect the coordinator to take EIU on board all through the school. We have a school statement and we have work placements but not yet for everybody. We do more in the lower sixth with the Crest Award and we still are involved in Project Trident. The UBI (Understanding British Industry) man has arranged placements for some of our teachers but it's a bit of a lottery, much depends on where they go and how clued up the place is. One of the staff got most out of what she called a focused placement about women in management – she happens to be our equal opps teacher. She made some good contacts for careers interviews so EIU and careers education often come together at Key Stage 4.

Q: How do you find time for EIU? Do you have a regular slot?

H: Well, I'm not the expert on the timetable but most of what goes for EIU in the lower school is done in time originally allocated to personal and social education. If we have an Industry Day we organise it in advance and cancel all other activities. We've had two since I've been here. We get a lot of help from our 'tame' industrialists. A couple of them – I think they're young managers – put in hours beyond the call of duty planning and working with groups of teachers and pupils. They're great – I think that we are one of the schools that their particular company has adopted. The teachers appreciate it. We've also had a Baker Day on cross-curricular themes and one of the options was EIU – it was a popular group with the staff and many of the primary teachers joined that group so we have some cross-phase cooperation. I'd like to see that go further. Your teacher (reference to an INSET Fellow for schools/industry links who works from the Department of Education at the University of Liverpool) came to that day. She's very lively – she was convincing – she got us going. Great.

Q: What about time for EIU in the upper school and the sixth form?

H: Yes, I was forgetting your question and rabitting on – do stop me. . . . We still use PSE time intermittently or perhaps spasmodically is a better word. It's not satisfactory but what can we do? We still have an options system. When everybody has to be timetabled for everything in the NC I'm not sure how we will cope. I think the cross-curricular themes will suffer. Any professional – head – who knows the job will find ways of justifying to the governors that the dimensions and themes are being adequately covered. I expect that environmental education will get lost in geography and science. Careers education and EIU will merge into each other with more emphasis on careers as they progress through the school but it will all have to come out of PSE in a much more systematic and structured way. That's why I prefer the term spasmodic – you can't be doing all these things all the time – that's out of the question. Systematic coverage in concentrated, planned periods is best. Don't ask me about citizenship, civics, politics or whatever the vogue term is – just don't know. We can't squeeze any more out of PSE but perhaps history or humanities will give some ground – or perhaps they won't – but you're not asking me about that are you? We'll have to find ways of getting staff to cooperate who don't wish to. Now, there's a problem.

*Q:*We'll have to stop there. Thanks for your time.

H: All that I can say is thanks for the opportunity – we don't make time to think out loud with somebody who is prepared to listen. I'll learn a great deal from going through the transcript.

The above is an example of an experienced reflective practitioner in action. He claimed not to have read much of the literature produced by the NCC but he did have a copy of Curriculum Guidance 4 (NCC, 1990b) on his desk. He began tentatively and slowly expanded his ideas. In essence his solution to the curriculum overcrowding problem is pragmatic. If he is required to produce a broad and balanced curriculum which follows statutory guidelines then themes will get systematic but sporadic treatment and his supporting documentation and skills in communication will be used to justify practice to the governors and presumably to visiting inspectors and parents. The basis of his professional judgements and practice will have to be made more explicit than hitherto but by implication he sees the changes as adjustments rather than a major overhaul. He also hints that a major problem in implementation will be one of staff development which will mean more professional collaboration in both planning and teaching than is the case at the moment in this school. If his attitude in general is the same as that conveyed during the interview then there will be no sense of panic engendered as they cope with multiple innovation but much more of a careful, reasoned approach about how best to utilise those things that are in short supply in all schools, namely time, money and people.

The next brief example is from an interview with the head of a medium sized 11 to 16 comprehensive secondary school which exists in an LEA alongside a selective grammar school. The school is situated on the edge of a very large industrial complex and over the years, from its early involvement in the first wave of TVEI, the school has developed positive links with many local firms and with the local careers service and until recently was well supported by LEA advisory staff. The head is very proud of the 200-plus opportunities for placements that he is able to call upon. The school has a deputy head responsible overall for outside links, has a TVE coordinator responsible for work related to that initiative and a separate teacher in charge of careers. Because of changes in the roles of the local advisory service the LEA has encouraged the development of a business/education partnership which is funded partly by the local Training and Enterprise Council (TEC) and partly by a large industrial multi-national company which has a local branch. The partnership is directed by a seconded industrialist and by a seconded teacher. At the moment the partnership works in four schools and is responsible, through the TEC, for coordinating all local work placements for pupils and for teachers and for supporting in-school activities which encourage EIU. The partnership is also responsible with the local Careers Service for the promotion of a Compact scheme which does not guarantee pupils from the schools a job but does guarantee them an interview at which they are considered for employment.

TECs were launched in the spring of 1989 to attempt to coordinate what to some people is a plethora of schools/industry initiatives and to help schools and colleges to smooth the transition from education and training to work. Eventually there will be a network of 100 TECs but in many areas they have been slow to get off the ground. Two-thirds of the membership of each TEC will be from local industry and the rest will come from local government councils, community groups, trade unions and education. TECs are run as limited companies. In launching the initiative, Margaret Thatcher declared to a gathering of industrialists: 'the Government is handing over to you', thus indicating that she thought that education had in this respect failed to meet the needs of individuals and those of industry and that industry had the answers to the education and training problem and with their enterprise they could deliver.

There is little doubt that there are many initiatives which confuse both schools and industry with the subtleties of difference and that it makes sense to attempt to coordinate and rationalise these. Any attempts to do this, however, must be sensitive to the differences between and within schools and colleges and to their institutional sense of autonomy. The degree to which this sensitivity can operate to the detriment of an initiative can be seen in the responses of the second head teacher when questioned about the Business/Education Partnership.

Q. Is the partnership working?

H: The general direction of the partnership is not as I perceived it and I am trying to get the best out of it without being certain that those people can even offer the best. . . . I actually perceive that there's a lot of political fighting, that somehow, what operates now is the direct line to the TEC so the LEA has almost been bypassed . . . there's also the problem of where careers sits in all this . . . they're turning it all into careers. . . . I have some sympathy with them but this head-teacher is very reluctant to put his 200 work placements into the pot to benefit others without knowing what he is going to gain from the system . . . at the moment I appear to be giving more than I am actually receiving back.

Three points can be made from this short extract which seem to have some general significance. First, if the TECs are seen as outsiders who know little about the workings of schools, mistakes are going to be made in the ways in which schools are approached and unfounded assumptions are going to be made about how things happen in schools. In particular, if a school has had former successful links with the LEA and suddenly the authority seems to have abdicated its interest in schools/industry work, this disturbs the traditional balance of power and affects the working of an initiative. Second, particularly at Key Stage 4, the links between careers education, the Careers Service and EIU need to be thought through. In this particular case, the assumptions from the TEC and from the directors of the partnership were that EIU is synonymous with making decisions about careers. In general it raises the question about whether careers education is similar in kind to the other National Curriculum cross-curricular themes. Finally, if a coordinated local initiative is perceived as offering a less adequate service than the one the school has worked out on its own then cooperation and collaboration is only going to be, at best, partial and reluctant. This head justifies his possessiveness over his large network of placements by saying that his pupils will suffer if he shares these with other schools. Such a deeply embedded and understandable stance is difficult to shift./

An attempt has been made to provide some convincing evidence from the examples above that EIU is having to fight for a place in the curriculum, especially at Key Stage 4, and that there is a tendency for it to merge with and become indistinguishable from careers education. If this is becoming the case pre-16, what is happening to attempts to provide education for EIU beyond the age of compulsory schooling? Is, for example, EIU able to find a place in the curriculum of A-level students who are mainly on the academic trajectory to higher education? Here it has to be seen as not competing for students' time and for resources in the general pursuit for entry grades set by admissions tutors. TVEI was envisaged as a four-year course from age 14 and was intended to make an impact post-16. What has this been?

As indicated above, there is no statutory entitlement to a National Curriculum above the age of 16. At this stage, EIU has to fight for its place alongside the public examinations of A and AS levels, initiatives from awarding bodies such as BTEC and the development of General National Vocational Qualifications (GNVQ). Recent thinking about how the latter may develop provides the possibility of an alternative pathway to further and higher education through a system of units of credit which can overlap and become equivalent to AS level in an attempt to give technical and vocational education more equality of recognition in comparison to 'academic' examinations. This system, if implemented, would also have the effect of broadening the traditional sixth-form curriculum. However, under the political climate which prevails at the beginning of the 1990s, this flexibility is hardly likely to be allowed to change in any radical way existing A-level syllabuses. The core skills identified by the NCC working party and laid out in the early part of this chapter could be incorporated into post-16 programmes but for those students on the academic pathway to higher education, time for such activities would probably have to be taken out of current General Studies time and any accreditation associated with the programme would be interpreted as 'value-added' and included in a Record of Achievement.

A pilot study to test some of these ideas was carried out in a project based at the University of Liverpool from 1987 to 1992. The 'TVEI 16–18 Enrichment Project' was an attempt to motivate able students to partake in TVE activities and to link these activities with higher education.

The enrichment programme is optional; students take it only after counselling from tutors. The work is conducted in small groups of four or five and during the pilot phase, teachers were not allowed to take on responsibility for more than two groups. The groups are expected to spend no more than two-and-a-half hours per week on their projects. Individual projects within the programme are planned, designed and resourced by small teams of teachers, industrialists and members of faculty from the university. The process skills which correlate closely with the core skills 16–19 of the NCC were developed through discussions with industrialists, academics and teachers. In summary, these are:

- setting targets and working to deadlines, including reviewing progress;
- developing interactive skills for group work and for the organisation and management of group decision-making;
- information handling and processing skills from traditional library skills to the use of IT; and
- communication skills for use in a variety of contexts with different audiences.

A full evaluation of the programme is being undertaken. To date, this has provided evidence that EIU can be delivered with considerable understanding to these able groups of students and that the project work enhances and enriches their traditional academic courses and does not interfere with them. The programme also provides training in the core skills and establishes a platform for genuine cross-curricular activity. At the same time the programme demands a particular teaching style from those involved. The adults are expected to take on the role of being a resource to the students and acting as consultants. The role demands standing back from the action and only intervening when asked to do so. Naturally, this uncovers a considerable training need for many of the teachers involved (Derricott, 1990).

The enrichment project indicates ways in which the core skills identified by the NCC can be embedded in the practices of teachers preparing students for traditional A-level examinations. Clearly, work of this kind is meeting the criteria of EIU by developing valued skills and providing opportunities for working directly with industry. There are, however, major barriers in the way of a wider implementation of the project's ideas. During the enrichment pilot programme negotiations were opened with a major examinations board to agree an AS syllabus to support the approach being advocated. Despite considerable backing from industrialists and educationists, the proposal was turned down by SEAC who could not accommodate aspects of group assessment into their view of examinations at this level. It is possible that the pathways

via GNVQs will provide a compromise solution but at the time of writing the way in which the rhetoric of support for the approach advocated by the enrichment project clashes with the hard reality of dominant traditional educational values appears to be symbolic of the overall position of EIU, especially between the ages of 14 and 19. Advocates of EIU as a cross-curricular theme face a considerable struggle to implement appropriate practices into an educational system which espouses its values only on the surface.

NOTE

1. A personal competences model for use from 14 years upwards is being trialled in the contexts of schools, FE colleges, higher education and in industrial/service/commercial organisations.

REFERENCES

Baker, K (1987) 'The National Curriculum: Key to better standards', transcript of speech delivered by the Secretary of State, Manchester University, 17 September.

Derricott, R (1990), 'Orientation to education and training: a consideration of interventionist strategies to vocationalise the curriculum in the United Kingdom', *British Journal of Education and Work*, 3, 2, 71–90.

Further Education Unit (1989) *Towards a Framework for Curriculum Entitlement*, London: FEU.

Holt, M (1982) 'The great education robbery', *Times Education Supplement*, 3 December.

Johnathan, R (1985) 'The Manpower Services model of education', *Cambridge Journal of Education*, 13, 2, 3–10.

MacGregor, J (1990) Secretary of State's speech to the Society of Education Officers, 25 January.

Murray, P R S (1989) 'Managing an educational change: an evaluation of the implementation of TVEI in a local authority', PhD Thesis, University of Liverpool.

NCC (1990a) *Core Skills, 16–19: A Response to the Secretary of State*, York: NCC.

NCC (1990b) *Curriculum Guidance 4, Education for Economic and Industrial Understanding*, York: NCC.

Chapter 7

W(h)ither International Understanding?

Bill Marsden

THE NATIONAL CURRICULUM AND INTERNATIONAL UNDERSTANDING

This chapter has a two-fold purpose. The first is to argue that for all the meritorious work of the National Curriculum subject working groups, and of those contributing their expertise to the National Curriculum Council's cross-curricular guidance documents, an over-riding political voice has determined that the National Curriculum in England and Wales should not only be national in the sense of a common curriculum for the nation, but also 'nationalised'. This radical shift has served to marginalise the international dimension, not least in history, where from an early stage Kenneth Baker demanded a stronger infusion of British history. Other histories (interestingly, those of Ancient Greece and the Roman Empire being the exceptions) were consigned, as far as Key Stage 4, to the periphery, as supplementary rather than core study units. An equivalent marginalisation of education for international understanding has occurred, as we shall find, in the official cross-curricular guidance documents. By contrast, the Statutory Orders on modern foreign languages offers in the 'Areas of Experience' section of the programme of study, an explicit proposal on 'The International World'. The potential value of this is of course constrained by its absence in the primary phase, but in the secondary phase the pursuit of international issues through the 'target language' is a potentially important contribution.

The second purpose of this chapter is to suggest that, notwithstanding the discouraging regress, education for international understanding need not wither if teachers refer back to an ecumenical, recent and relevant range of good practice. While the exigencies of the situation indicate that for this to happen a key role must be played by permeation through the geography curriculum, it is argued that in principle this is in harmony with strongly-founded traditions of progressive, enquiry- and issues-based geography teaching, and that geography provides a necessary though not in itself sufficient base for the realisation of a broadly-conceived education for international understanding.

In its reponse to a challenge from Sir Keith Joseph, then Secretary of State for Education, to demonstrate the distinctive contribution geography could make to a balanced coverage of controversial issues and to political understanding, the Geographical Association stressed again and again the subject's various offerings to education for world citizenship and international understanding. A similar focus was of course also present in the agendas of those running projects or courses in world studies, global studies, development education and peace education. Experts from these fields regularly contributed to publications on international understanding produced by geographers and vice versa. The Council for Education in World Citizenship argued in 1987 that geography was the most obvious subject for the promotion of education in international understanding. Whatever the mode of delivery, by the time of the Education Reform Act in 1988 there was a broad consensus that this was a key cross-curricular area. The label itself was loose, as indeed were the other labels in a constellation of titles. In essence, international understanding has knowledge and understanding *and* values and attitudes components. It is involved at the global level with developing understanding of a wide range of often controversial place-specific social, environmental, economic, political and cultural issues. It is involved with fostering an emphathetic understanding of other people's views, ways of life and social and cultural perspectives. It is involved with the interaction of peoples and their environments on local, regional, national and international scales.

Despite a brief to the National Curriculum geography working group to ensure that pupils develop an informed appreciation and understanding of the world in which they live and the physical, economic, political and cultural ties that link peoples living in different parts of the world, an address to education for international understanding seems after 1989 to have disappeared from the official charts. Unlike environmental education, which became an attainment target (as environmental geography) within geography and an officially recognised cross-curricular theme, international understanding, while not ignored, was not given similar status to that obviously important, but arguably no more important, area, and was peripheralised in the NCC's guidance documents on cross-curricular themes. The evidence from geography, history, education for economic and industrial understanding, health education, education for citizenship and environmental education is consistent with the view that the government wanted a nationalist curriculum, giving priority to the transmission of traditionally English cultural values, notwithstanding its range of commitments to international organisations, and its obligations to a multi-cultural society, contenting itself with some token support for educational inputs at the international level.

The statutory orders for geography, and international understanding

In its choice of personnel and in the terms of reference laid down to the geography working group, it was clear from the start that the Secretary of State's intention was that geography should fairly and squarely be defined as the study of places. But which places? There was to come significant and worrying back-tracking on the primacy to be given to a genuinely global, issues-based approach to the study of distant places.

While there was criticism of the geography working group's choice of distant places to be studied in its interim report, the initial statement of aims devoted an appropriate degree of attention to issues of global interdependence, care of the earth and its peoples and to promoting geographical enquiry. Stress was laid on links with certain cross-curricular themes, some of which, such as political education, international understanding and development education, were never heard of again from any official body. Notwithstanding the limitations, hardly surprising in the light of the unseemly haste with which the working group had to produce its report, it offered a more wide-ranging global agenda than anything to follow.

The broad aims of the interim report were reproduced almost verbatim in the final report. The main points of criticism on the place attainment targets were in general redressed. The subsequent NCC consultation report, while containing further refinements and improvements, watered down the place element in the geography curriculum from the three original ATs to one AT (2), defined as 'knowledge and understanding of places', while maintaining three thematic ATs (2), (3) and (4).

While the promotional rhetoric of geography as a subject uniquely qualified to foster international understanding was not satisfactorily implemented, at least the consultation document retained, especially in AT5, a residual issues-based orientation and a recognition of the relevance of an attitudinal dimension within such issues. Within a week of the belated receipt of the consultation report (delayed at the printers), the draft orders appeared, a classic case of an attenuated gestation (for which, read consultation) period and premature birth. The third

Secretary of State to emerge during the time of the deliberations of the working group and the emendations of the NCC had a distinctly Thatcherite view of the value of discussion and crisper concepts than his predecessors of what was required. He put right out of court the idea that there should be any appraisal of pupils' exploration of values and attitudes through the statements of attainment. The specification of localities to be studied was further weakened, the European dimension almost disappearing from the primary phase. The statutory orders for geography thus represent in some degree a victory for reaction but, fortunately, in no essential sense subvert the potential of the subject to remain a central force in the delivery of education for international understanding.

Cross-curricular guidance and international understanding

If the statutory orders for geography are now somewhat pruned of content relevant to education for international understanding, particularly in the all-important primary phase, their peripheralisation in the cross-curricular NCC guidance documents represents a more blatant politicisation of the curriculum. As a group these documents are unexceptional: attractively presented, clearly-structured, easy to follow and well-balanced according to all the dictates of the 1986 and 1988 Education Acts. They are, however, disturbingly Anglo-centric in tone.

The absence of education for international understanding, or some equivalent in Curriculum Guidance 3 on the whole curriculum self-evidently set the seal on what was to follow. Of all the contenders from which the cross-curricular dimensions and themes were selected, global education, or some variant, was the one most conspicuous by its absence in the final solution.

Curriculum Guidance 4 on education for economic and industrial understanding (EIU) contains undoubted strengths, based as it is on authentic conceptual underpinnings drawn from previous development work in economics education. At the same time, its central preoccupation is with EIU as a means of address to indigenous problems. The very title, economic and industrial understanding, suggests parochialism, and is a political rather than an academic construct. Obviously contact with industry can enhance industrial understanding and, if used in this way, all is well and good. But if its prime intent is to prop up future British industry and inculcate the values of enterprise, then the educational motivation is not the priority. The focus is not entirely narrow and rigid. There is mention of attitudes and respect for alternative economic viewpoints and of the impact of economic decisions on human rights, but in the practical and arguably more diagnostic illustrative detail the key emphases are shops and super-stores, workplaces and mini-enterprises in this country. Perhaps significantly, the only identification of a global dimension lies in the reference to the permeation of EIU in geography. In essence, however, as Carter (1991) has intimated:

> The balance is wrong in its under-emphasis on the global economy, the role of multi-national corporations, the ever-widening gap between rich and poor, and the knock-on implications for global management (p. 30).

Curriculum Guidance 5 on health education is, if anything, even more domicentric, and appears almost exclusively concerned with health hazards facing consumerist white English persons. There is no mention until Key Stage 4 of food shortages and surpluses and of malnutrition, and no clear specification even there that this is a world problem. While geography's contribution to a matrix of health education components: substance use and misuse; family life; safety; health-related exercise; personal hygiene; environmental aspects; and psychological aspects, is indicated, it is missing in the section on food and nutrition, the one most clearly associated with a global element. Regrettably this aspect is not specifically identified in the geography statutory orders until Key Stage 3. For younger children, therefore, the message is that health education is not at all about global interdependence and international understanding, but about their personal concerns and how friendly adult agencies are working to assist them. It would be perverse to play down the importance of the issues with which health education seeks to grapple, but the fact that children are to become citizens of a world the majority of whose members are in a morbid state of ill-being, as well as of a nation a minority of whose members pay a mortal price for western consumerist life-style excesses, appears not to have been recognised. The document surely does not offer a broad and balanced curriculum in this area.

By contrast, both in the statutory orders for Geography and in Curriculum Guidance 7 on environmental education, the heady ambience of a new secular religion is all too apparent. Here issues of the ozone layer, acid rain, the destruction of natural habitats and similarities and differences between peoples in the way they use their environments, have a genuinely global feel. Again, much of the delivery on the global scale would appear to be left to geography. Most of the exemplification offered is, however, Eurocentric, though there is an interesting exception in the suggestion of a Key Stage 2 approach to international links through a case study of twinning with a Nigerian school, a truly global approach to be discussed later.

In the context of education for international understanding, the highest expectation and most disappointment must be linked with Curriculum Guidance 8 on citizenship education. In a spectrum of definitions of citizenship education which stretches between an anachronistic elementary school-type civics and a radical world citizenship approach to the problems facing human-kind around the millennium, the document is manifestly more comfortable in the former domain. Machon (1991) notes that the subject of citizenship education is not here associated with an energetic and contested debate, and is more about 'subjects' than 'citizens'. Many of the right words: a pluralist society, equal opportunities, global issues, the need for international cooperation, are mentioned in the overall rhetoric. In terms of delivery, however, there can be no question that the document is rooted in a Victorian conception of citizenship, as evidenced in the practical suggestions offered. Thus 'progression' is demonstrated by how to cover 'the citizen and the law'. The key concepts of rules and laws loom large. Geography is given some small responsibility for introducing a global element, as in a section on pluralism but, as Machon concludes, the prevailing tone is 'as British as a semi-detached house in the suburbs' (p. 128). Perhaps the give-away section is the old-time call of the civics syllabus to enlist the help of adults in work roles in the community by inviting them into the classroom to explain their work. While the idea is pedagogically sound, the choice of roles has an obvious hidden agenda: the organisations recommended for use are the St John's Ambulance, the Red Cross, the Scouts and Guides, the Duke of Edinburgh Awards Scheme, Outward Bound, Operation Releigh and, 'of the greatest importance', the Police Service, which 'can help in developing the ethos of a school' and 'support active, participative citizenship through enterprises such as Junior Crime Prevention Panels' (NCC, 1990). Whither Oxfam, VSO, UNESCO or, dare one mention, Amnesty International?

It is surely an evasion of responsibility, as a country in membership of international organisations and subscribing to their mission statements, that in a National Curriculum document on citizenship education references to UNO, to take one example, are limited to its role in 'work, employment and national and international economics', and 'its main features' as an international organisation. There is also a suggestion of covering major conventions on human rights and of comparing the situation in Britain on racial issues with that in other countries. Unlike aspects of perceived national significance, however, these matters of global import appear as marginal to the main thrust. So far as the National Curriculum cross-curricular guidance is concerned, therefore, and not least in the citizenship education document, we appear to have a situation of allowing international understanding to wither on the vine.

WHITHER INTERNATIONAL UNDERSTANDING?

While only the rank optimist would expect an easy future ride for education for international understanding, there are some grounds for hope. One lies in the dual economy of the National Curriculum, that is, the possibility it offers for a fruitful blurring of the distinctions between subject and cross-curricular approaches. Much ideological blood, sweat and tears have been expended on fossilising and polarising these distinctions. In practice, the judgement which has to be, and indeed has been, made is whether an issues-based/subject-permeated curriculum is more or less viable than a subject-based/issues-permeated one. It would seem it is too finely tuned a judgement to be comfortable with in this present divisive age.

As relevant contributors to the Dufour (1990) collection of papers on cross-curricular issues indicate, genuine international understanding can only effectively be achieved through activity-based educational procedures, for indoctrination is as of principle a narrowing and dehumanising process. What they fail to recognise, it would seem, is that the commitment to 'active student-centred learning and a focus on issues' is not merely a characteristic of peace education and most of the other themes considered in the book, but is also an essential diagnostic feature of good teaching in geography and other subjects. The claim that areas such as global education do not fit well in a subject-bound curriculum may have some theoretical purchase, but whether they do so or not depends on the nature of the subject teaching currently in practice, and not on stereotypical definitions of what it used to be (5).

In this context a dual economy might have the further benefit of defusing the fear said to be apparent in the public mind that education in international understanding is by its nature potentially indoctrinatory, with peace education looming large in the far-right demonology. It can therefore be argued that for those for whom the cause of education for international understanding has a high priority, in the frame of the National Curriculum can be found pathways which can be turned to advantage. One part of the delivery of global issues must, by broad consent, be via permeation through geography. At the same time, geographers would be foolhardy to claim that they can fully cover these complex issues without considerable cross-curricular support.

Where then, can global issues optimally be covered, bearing in mind the practical constraints of fitting subjects and cross-curricular dimensions and themes into scarce time-slots? It is apparent from the cross-curricular guidance, as well as from the general literature, that there is an enormous overlap in areas broadly covered by such terms as 'integrated humanities' or 'integrated social studies' and, more specifically, by global education, development education, world studies and the like. The need for some over-arching mode of delivery is apparent, accompanied by some careful auditing, not least of global coverage, to complement the geographical input.

On the basis of the recommendations of the official documents relating to the National Curriculum, however, and bearing in mind the 'by definition' focus of a humanistic place-oriented geography on world study, it would seem that it must remain logistically and, arguably, legitimately so, the major vehicle for delivery of education for international understanding.

Apart from the nature of its content, another reason for seeing geography in this role is that inevitably the subject label is more 'legible', as Storm (1989, p. 293) has put it, to parents than the idiosyncratic cross-curricular titles used to cover different and at times the same aspects of international understanding. To the converted, all the categories in, for example, Dufour's (1990) collection on cross-curricular approaches to social education: pre-vocational and vocational education; personal and social education; health education; media education; peace education; gender education; multi-cultural and anti-racist education; global education; environmental education; trade union education in schools; and human rights education, have some purchase, albeit representing an unfeasible feast for any curriculum coordinator to accommodate in the timetable if separate slots are being claimed. But they may appear opaque or subversive to the uninitiated.

In setting free education for international understanding from the parochial cage that represents the official frame for the National Curriculum, the key is held by teachers committed to what is for them a moral responsibility and for their charges a cultural entitlement. Fortunately there is much recent good practice to call upon, and potential space for delivery by percolation into the bedding planes and joints of what are the highly permeable strata of the National Curriculum.

Four examples of such good practice are here called upon, deliberately chosen to indicate that while the vehicle for delivery may often have to be in the geography slot on the time-table, the approach should be broad and integrated. Three of the examples are listed at the end of this article as part of a larger series of helpful materials.

The first example of good practice is taken from Fisher and Hicks'

World Studies 8-13: A Teacher's Handbook (1985), an outcome of a curriculum development project initiated jointly by the Schools Council and the Rowntree Trust. Among its many constructive features it asks teachers to question their own values, attitudes and approaches to enquiry learning, crucially demonstrating that good content, good pedagogy and good social education must be connected in any credible agenda for promoting education for international understanding. Not least of value is the synthesis of good practice which *World Studies 8-13* offers, framing its objectives, for example, on an earlier World Studies project led by Robin Richardson (see the further reading list at the end of this chapter).

- *Knowledge objectives*
  Ourselves and others
  Rich and poor
  Peace and conflict
  Our environment
  The world tomorrow
- *Attitudes*
  Human dignity
  Curiosity
  Appreciation of other cultures
  Empathy
  Justice and fairness
- *Skills*
  Enquiry
  Communications skills
  Grasping concepts
  Critical thinking
  Political skills

(Fisher and Hicks, 1985, p. 25).

Such a list makes clear the potential limitations of a narrowly-focused geography in fostering education for international understanding. At the same time, there is a similar limitation in the Fisher and Hicks approach in the absence of address to vivid space-specific knowledge, particularly important for the age phase for which they were writing. At the same time, as Fisher and Hicks indicate, there is no innate conflict between distinctive subject offerings to world studies and their own cross-curricular approach (Fisher and Hicks, 1985, pp. 22-3).

A particularly valuable recent publication is the Development Education Centre's *Theme Work: A Global Perspective on the Primary Curriculum in the '90s*. This offers a more dynamic and internationalist approach to cross-curricular themes and dimensions than anything in the NCC's cross-curricular guidance. It critically demonstrates that children themselves have to learn to interact cooperatively in a group context and empathise with their peers before they can effectively do so with groups and peoples beyond their immediate ken. At the same time, the approach provokes testing questions, probing, for example, what is meant by a 'country', clarifying definitions, building up content about the home country, creating an imaginary country, developing activities to deepen and share understandings about life in different countries, exploring the nature of trading links and interdependence, as well as addressing issues of conflict and the consequences of unequal power and opportunity (DEC, 1991, pp. 49–64). While there is a stronger place-specific quality than in Fisher and Hicks, the stress on general global issues contrasts with an under-representation of specific places within the book, though some reference is made to sources of detailed case studies of people in actual locations. The links with geography are strong, if often left implicit.

The third example of good practice is that offered in Beddis and Mares' *School Links International: A New Approach to Primary School Linking Round the World*, the result of a joint venture between Avon County Council and the Tidy Britain Group. This too represented a cross-curricular approach but one more place-specific and real, allowing the wider world to be brought into the classroom through the linkage of classes in schools on a global basis. Contacts with nearly forty countries were instanced (Beddis and Mares, 1988, p. 46). The book specifies in detail the aims, outcomes and procedures to be followed in establishing such links. Before his untimely death, Rex Beddis was well-known as a geographer and as director of the Schools Council Avery Hill geography project, which pioneered approaches to welfare geography transcending previous narrow boundary definitions, bringing it fairly and squarely into the cross-curricular domain of controversial and values issues. To secondary geography teachers, Beddis's work in this sphere is recognised in textbooks such as *The Third World: Development and Interdependence* (Beddis, 1989), one of a large number of models of good practice in this area eschewing narrow disciplinary confines, though written by geographers. The School Links International initiative was also formative in anticipating two features of the National Curriculum geography working group's recommendations, in promoting detailed locality studies, not only of the local area, but also of distant places, and in subsequent recommendations for twinning arrangements, as a means of establishing links with schools in other localities in this country and abroad.

One of the valuable features of the work of the Development Education Centre and similar agencies, and found also in Fisher and

Hicks (1985), is the production of checklists offering criteria for good practice in global studies, whether implemented through geography or in some cross-curricular way.

An example of such a checklist is taken from an article by the geographer Michael Storm (1991). He pointedly draws attention to the anomaly that the marginalisation of the wider world dimension is no new discovery of secretaries of state for education but was endemic in so-called progressive primary school practice, the experiential learning ideology tending to confine geographical input to studies of the local area. In this article Storm, following other geographical writers, suggests in embarking on locality studies the need to avoid 'the hazards of encyclopaedism, exoticism, anglocentricity, the unremittingly pathological' (pp. 20–22), and to use a geographical frame of reference embodied within the following key questions:

- Where is this place?
- What is this place like?
- Why is this place as it is?
- How is this place connected to other places?
- How is this place changing?
- What would it feel like to be in this place?

Address to such questions requires an exacting but necessary and worthwhile process of collecting a wide and vivid range of resources, defined by Storm as 'optimum ingredients':

- a specific, named location
- named people, preferably families
- a focus upon the lives of children
- pictures of people, landscapes, buildings, etc.
- maps and plans, from small-scale contextual locating maps to large-scale maps and plans
- the pattern of daily life – 24-hour time-lines
- how income is earned, or subsistence organised
- diet, clothing, housing details
- patterns of movement: work, school, trade
- shopping, market, trade activities
- leisure activities: festivities, special occasions
- connections with the wider world through travel, etc.
- changes in landscape and lifestyles
- aspirations and problems
- data on climate, descriptions of weather
- descriptive or imaginative literature set in the locality or its region.

Such an optimal collection clearly offers wide-ranging opportunities for creative and problem-solving activities, as well as promoting a knowledge and basic understanding of distant places, as Storm again outlines (pp. 20–22):

- planning a journey to the locality
- making posters to 'advertise' the locality
- making pictures of the locality as envisaged
- writing a letter home from an imagined visit to the locality
- writing to a child who lives in the locality
- writing a story set in the locality
- making a map of the locality and its setting
- making a model of the locality or a building within it
- making diagrams (time-lines) of the daily rhythm of life
- making calendar diagrams of the annual rhythm of life
- making climate graphs for the locality
- debating a proposed change in the locality
- writing a letter to an envisaged local newspaper or enquiry
- collecting or sorting other material about the country in which the locality is situated: postcards, pictures, stamps, coins, flags, product labels, etc.

SUMMARY

One of the regrettable features of the NCC's guidance documents is how narrowly 'national' they all seem. The development of a global social, economic and political literacy is a morally and culturally prescriptive contribution to international understanding, and is demanded by official global organisations to which the British government subscribes.

The need is for worldwide agreement on an approach to education for international understanding which can differentially be pursued according to national wishes, though within a common and humanistic values framework. The International Geographical Union has, for example, made a start on this in its promotion of an 'International Charter' on geographical education. Central to the aims is that:

Education should be infused with the aims and purposes set forth in the Charter of the United Nations, the Constitution of Unesco, and the Universal Declaration of Human Rights, particularly Article 26, Paragraph 2, of the last-named, which states: 'Education shall be directed to the full development of the human personality and to the strengthening of human rights and fundamental freedoms. It shall promote understanding, tolerance and friendship among all nations, racial or religious groups, and shall further the activities of the United Nations for the maintenance of peace' (Commission on Geographical Education, 1991).

The Council of Europe's Recommendation No. R(85)7 on the 'Teaching and learning about human rights in schools' among other things encourages schools to observe International Human Rights Day (10 December). There would appear to be a more inhibited approach to human rights education in the report of the Speaker's Commission on Citizenship. In its evidence to the NCC it accepts that the main international charters and conventions to which Britain is a signatory should provide reference points within the classroom, but qualifies that with the view that Britain has different traditions and that the development of entitlements and responsibilities within the UK should be a feature. Indeed it has become a focus.

Whither, then, international understanding as part of the curricular entitlement of pupils? If indeed it is the current government's intent not to give priority to the international educational commitments it is already signatory to, then teachers, as already indicated, are saddled, or privileged to bear, the responsibility for seeing it does not wither. Let Confucius be their talisman, as favourably quoted 230 years ago by Oliver Goldsmith (1762):

> Let European travellers cross seas and deserts merely to measure the height of a mountain, to describe the cataract of a river, or tell the commodities which every country may produce; merchants or geographers, perhaps, may find profit by such discoveries, but what advantage can accrue to a philosopher from such accounts, who is desirous of understanding the human heart, who seeks to know the men of every country, who desires to discover those differences which result from climate, religion, education, prejudice and partiality . . . Confucius observes that it is the duty of the learned to unite society closely and to persuade men to become citizens of the world.

REFERENCES

Beddis, R (1989) *The Third World: Development and Interdependence*, Oxford: Oxford University Press.

Beddis, R and Mares, C (1988) *School Links International: A New Approach to Primary School Linking Round the World*, Avon County Council/World Wildlife Fund.

Carter, R (1991) 'A matter of values', *Teaching Geography*, 16.

Goldsmith, O (1762) reprinted in Friedman, A (ed.) (1960) *Collected Works of Oliver Goldsmith: Vol. II, The Citizen of the World*, Oxford: Clarendon Press.

Commission on Geographical Education (1991) *International Charter on Geographi-*

cal Education, International Geographical Union Commission on Geographical Education, Newsletter 22.

Development Education Centre (1991) *Theme Work: A Global Perspective in the Primary Curriculum in the '90s*, Birmingham: Development Education Centre.

Dufour, B (ed.) (1990) *The New Social Curriculum: A Guide to Cross-curricular Issues*, Cambridge: Cambridge University Press.

Fisher, S and Hicks, D F (1985) *World Studies 8–13: A Teacher's Handbook*, Edinburgh: Oliver and Boyd.

Machon, P (1991) 'Subject or citizen?', *Teaching Geography*, 16.

NCC (1990) *Curriculum Guidance 8: Citizenship Education*, York: National Curriculum Council.

Storm, M (1989) Geography in schools: the state of the art', *Geography*, 74.

Storm, M (19??) 'Worlds elsewhere?', in *Primary Geography Matters*, Sheffield: Geographical Association.

FURTHER READING

The following texts and handbooks are all useful in offering advice and plans of campaign for introducing a global dimension into the curriculum. Some are written by geographers and by social scientists and members of cross-curricular projects on world studies, global education, development education and the like. It is often difficult to tell which is which.

Batty, P (1991) *Global Connections: A Review of British Secondary North–South International School Links*, Isleworth: British and Foreign Schools Society, West London Institute of Higher Education.

Bale, J (ed.) (1983) *The Third World: Issues and Approaches*, Sheffield: Geographical Association.

Centre for World Development Education (1989) *The Primary School in a Changing World*, London: CWDE.

Daniels, A and Sinclair, S (eds) (1985) *People Before Places?: Development Education as an Approach to Geography*, Birmingham: Development Education Centre.

DES (1986) *Curriculum Matters 7, Geography from 5 to 16*; London: HMSO.

Fien, J and Gerber, R (eds) (1988) *Teaching Geography for a Better World*, Edinburgh: Oliver and Boyd.

Heater, D (1980) *World Studies: Education for International Understanding in Britain* London: Harrap.

Hicks, D and Steiner, M (1989) *Making Global Connections: A World Studies Workbook*, Edinburgh: Oliver and Boyd.

Hicks, D and Townley, C (eds) (1982) *Teaching World Studies: an Introduction to Global Perspectives on the Curriculum*, London: Longman.

Pike, G and Selby, D (1988) *Global Teacher, Global Learner*, London: Hodder and Stoughton.

Richardson, R (1980) *Ideas into Action: Curriculum for a Changing World*, London: One World Action Trust.

Selby, D (1987) *Human Rights*, Cambridge: Cambridge University Press.

Stumpe, L H (1988) *School in the World: Teaching for a Global Perspective*, Merseyside Association for World Development Education, available from CWDE, London.

Walford, R (ed.) (1985) *Geographical Education for a Multi-cultural Society*, Sheffield: Geographical Association.

The Centre for World Development Education (Regents College, Inner Circle, Regents Park, London NW1 4NS) is a useful source for some of these publications. The various aid agencies are increasingly producing 'locality study' materials on the economically developing world. The stress on detailed locality studies at home and abroad is indeed one of the most constructive contributions to education for international understanding in the National Curriculum programmes of study.

Chapter 8

Citizenship in the National Curriculum – A Struggle for Survival

Janet Strivens and Cliff Jones

Since the National Curriculum Council announced that 'education for citizenship' was to be one of its five cross-curricular themes there has been an outpouring of literature analysing every aspect of the subject. There have been numerous attempts to define and redefine the concept and much analysis of why this particular term has been chosen at this historical juncture by the powers that be. This chapter will not be centrally concerned with the definition of citizenship, which is excellently done elsewhere (Brown, 1990b, 1991; Foster and Kelly, 1991 and Heater, 1990; 1991).[1] Suffice it to say that there are good reasons why citizenship is a difficult concept to define. It relates to the dilemma all human societies have faced: that while human beings generally appear to need some degree of freedom of choice to thrive, they also need to live together and thus accept restrictions. History shows us the range and variety of attempts which different societies have made to resolve this dilemma, and their consequences, both short- and long-term. 'Citizenship' describes a set of these attempts with certain principles in common, though their different historical contexts bring about different realisations. For the teacher facing the responsibility of educating for citizenship, no simple definition can substitute for some awareness and understanding of the philosophical and historical debate.

This requirement that the teachers themselves have an educated understanding of the concept is the first factor in good education for citizenship. The main purpose of this chapter is to explore the other ingredients believed to be necessary; then we may dare to speculate on the likelihood that the new generation of school students will get this basic entitlement – the learning experiences that will develop the knowledge, critical understanding, skills and values of an active citizen.

THE PRESENT CLIMATE

It is vital to start by spelling out the features of the climate – educational, political and economic – in which the National Curriculum is now being operationalised. Economically, Britain is fighting to maintain a place among the 'first world' trading nations. Its traditional competitors in Europe, the United States and Japan are being supplemented and overtaken by the dramatic success of the 'Pacific Rim' economies. Money for public services is in short supply, with most of them perceiving themselves as underfunded. The education service has been among the most vociferous in drawing attention to the inadequacy of its equipment and the low morale of its workforce. In 1989 three publications (Confederation of British Industry, 1989; Trades Union Congress, 1989; Training Agency, 1989) showed a remarkable degree of consensus between both sides of industry, and with independent analysts, about the skills required in the workforce and the investment in education and training required to compete successfully into the new millennium.

Politically, we have had a decade in which the party in power has felt, most of the time, secure enough to carry through radical policies without too much concern for widespread consultation outside its own supporters or for 'minority' views. A Prime Minister who claimed that society does not exist was finally replaced by one who claims to dream of a classless society. The issue which brought this change about, and which has been rumbling ever louder throughout the decade, relates directly to definitions of citizenship; it is the issue of this country's relationship with the European Community and the value placed on 'national sovereignty'.

The focus on citizenship as an energising concept sharpened in 1988 when Douglas Hurd, then Home Secretary, took *active citizenship* as the theme of a major speech. Heater's incisive analysis (1990, pp. 295–304) shows how all the main political parties have claimed and used the term since then to strengthen their particular message. A Commission on Citizenship was set up in December 1988, chaired by the Speaker of the House of Commons, Bernard Weatherill, who appointed its members. Its report, published in 1990 shortly before the NCC's *Curriculum Guidance 8: Education for Citizenship* (1990), lays heavy emphasis on citizenship as voluntary service to the community. The foreword to *Curriculum Guidance 8* acknowledges that they 'have taken due account of the work of the Speaker's Commission on Citizenship'.

Education has had its own radical 'reforms' embodied in the 1988 Education Reform Act, amongst which the National Curriculum and its attendant assessment procedures have probably gained the highest profile. It is important to remember that at the start of the decade the Government was dealing with a highly devolved system and has had to put much of its energy into centralising its control. The 1988 Act is likely to succeed in radically changing the organisation and administration of the system. However, changing curriculum practice through legislation is considerably more problematic. To begin to understand the impact of the National Curriculum at the end of the 1980s we need to look back at some critical events and changes that have taken place in the 15 years since Callaghan's Ruskin College speech in October 1976. That speech initiated the 'Great Debate' with its theme of a closer relationship between schools and industry. Meanwhile, the Manpower Services Commission,[2] funded through the Department of Employment, was growing and spreading its influence at an alarming rate, ultimately encroaching on the Department of Education and Science's hallowed territory with the establishment of the Technical and Vocational Education Initiative (TVEI). Educationists' initial concern at this territorial invasion was allayed when it became clear that schools were using the relatively lavish funding to develop exciting courses involving problem-solving, teamwork and negotiation requiring new, more flexible forms of assessment and developing new relationships between students, teachers and adults from the 'real' world.

HMI published the first of the three 'Red Books', *Curriculum 11–16* in 1978. In setting out the eight 'areas of experience' which should constitute a common curriculum, the currency of skills and attitudes was firmly established alongside knowledge and understanding as valid curriculum objectives.[3] 'Red Book One' also picked up from the Great Debate the theme of preparation for adult life and the world of work. The third Red Book (HMI, 1983) introduced the key concept of an 'entitlement curriculum' for all students. Although the setting up of the National Curriculum by subject disciplines rather than areas of experience appeared to ignore the HMI's work, the focus on skills and attitudes and the language of entitlement were already embedded in educational thinking. It was necessary for the proponents of the National Curriculum to make the claim that this was indeed an 'entitlement' curriculum, a claim which has been seen as rhetorical by some, but which nevertheless exists as a lever for those who wish to press it.

Finally, there has been a quiet revolution in attitudes towards assessment during the last decade, which is now surfacing as a major battlefield. The change is illustrated both in the original planning for the GCSE criteria which insisted that students should be allowed to demonstrate what they knew, understood and could do, and encouraged coursework assessment; and in the major developments in profiling and Records of Achievement (partly stimulated by the need for more flexible forms of assessment for TVEI). The changes in attitude are summed up in the Task Group on Assessment and Testing report (DES, 1987) which was intended to provide clear principles for National Curriculum assessment. At the time of writing these intentions are increasingly being thwarted as assessment becomes more simplistic.[4]

All these phenomena have deeply influenced the climate in which the National Curriculum must now be put into practice. Areas of contention in the implementation of the National Curriculum, both in terms of content and assessment, bear witness to the tensions existing between at least three parties to the debate: teachers holding certain values about the aims of education; those with political power holding opinions about what constitutes appropriate education for the masses; and industry's more or less enlightened understanding of its own present and future needs as expressed through the programmes implemented and funded initially through the Department of Trade and Industry and now through the Department of Employment.[5]

THE CROSS-CURRICULAR THEMES

In terms of Government thinking, it is possible to see the cross-curricular elements of the National Curriculum as second thoughts, an attempt to mellow the rigidity of the discipline-based curriculum in the face of objections from both industry and education. It is important to remember that the cross-curricular elements, unlike the core subjects plus RE, are non-statutory, and the pressure on schools to accommodate the statutory subjects is intense. Secondary schools in particular need to have a strong faith, based on their past experience of the value of personal and social education courses, to create separate timetable space for the themes. Recommendations contained in *Curriculum Guidance 3: The Whole Curriculum* (CG3) (NCC, 1990) have been careful *not* to specify how the themes shall be tackled, putting the onus firmly on the schools to organise 'delivery'.

There are, quite predictably, large areas of overlap between the five curriculum guidance documents. Of the five, *Education for Citizenship* (CG8) (NCC, 1990) (the last to be published and also the shortest) has the most connections and could encompass most of the others. Again, this is not surprising since education for citizenship can be so broadly conceptualised that little falls outside it. It contains eight 'essential components':

- community;
- a Pluralist Society;

- being a Citizen;
- the Family;
- democracy in Action;
- the Citizen and the Law;
- work, Employment and Leisure;
- public Services.

Taken as a whole, this list of components and suggested activities reads much like the type of social studies course with which secondary schools are already familiar. Of the eight, three are distinctively in the area of politics: 'Being a Citizen', 'Democracy in Action' and 'The Citizen and the Law'. A teacher with no previous background or interest in the study of politics might particularly feel the lack of specialist knowledge in tackling these components. 'Community' and 'The Family' are key areas in sociology, figuring in all GCSE and A-level syllabuses. All eight components require a social science perspective to realise their potential in terms of knowledge and understanding. Interestingly, this is more explicitly recognised in the documents relating to some of the other themes (see Burrage, 1990).

It is certainly worth asking why the NCC went to the trouble of specifying five separate themes from the huge range of topics and areas which constituted personal and social education in schools. The choice was bound to appear either arbitrary or politically motivated. Perhaps the real effect of using the term 'Education for Citizenship' is to militate against the possibility that it will be offered as an examinable course. Citizenship is seen as a right, not a reward; teachers have the same reluctance to contemplate grading or even failing a student in 'citizenship' as they have in 'life skills'. A comment from the consultative conference called by the Speaker's Commission on Citizenship is noteworthy:

> Our consultative conference overwhelmingly supported the contribution of the Record of Achievement in recording and assessing a young person's citizenship *contribution* (emphasis added). The GCSE in citizenship was equally overwhelmingly rejected though some Commission members considered that it should not be ruled out altogether for those who wished to pursue it (Speaker's Commission on Citizenship, 11990, p. 38).

It is very likely that the ultimate challenge for the cross-curricular themes, which will determine how seriously schools are prepared to take them (that is, to resource them), is their accessibility, or rather, whether their assessment will count towards the school's position in the league tables of examination successes. Is the real battleground for citizenship in schools the value society (in the shape of parents and employers) is prepared to place on Records of Achievement? This question will be returned to in the final section. Let us first consider how a school which takes its responsibility for citizenship education seriously should approach the planning of its curriculum delivery.

DELIVERY SYSTEMS

Taking account of the pressures on timetables and teachers already commented on, it is particularly important for schools to reflect carefully about the nature of the delivery system they adopt. This section will briefly discuss a model of curriculum delivery and assessment designed to support 'education for citizenship' (taking this to be the most general of the cross-curricular themes). The features of such a delivery system should enable young people to demonstrate what they know, understand and can do; it should encourage (or indeed require) teachers to share the intentions of learning with students; and review outcomes of learning with students. Curriculum planning should now be taking place on a whole-school, departmental and individual basis and schools are being required to establish an assessment policy. The following curriculum planning and assessment model uses National Curriculum guidelines and also takes as central the features mentioned above.

The *first stage* requires teachers and departments to blend programmes of study laid down by the National Curriculum with the school statement of aims in order to create schemes of work adjusted to the ecology of the particular school. A school which makes no reference to citizenship at this point will make life difficult for a teacher attempting to introduce it into teaching and lesson plans in an individual, ad hoc or implicit basis.

The *second stage* of an effective delivery and assessment system is likely to be concerned with the drafting of teaching plans to cover, say, half a term. Clearly thought out teaching plans are the basis for teacher assessment. The components of such a plan provide a framework for lesson plans and activities and a sounding-board for the review of evidence and outcomes. No matter what changes are made to the relationship between teacher assessment and standard assessment tasks (mandatory and non-mandatory) there will always be a need to plan for teacher assessment as an assurance of quality.[6]

Teaching plans which will generate the evidence for good quality teacher assessment will:

- explain the circumstances in which learning will take place (in

Figure 8.1 *Creating an assessment policy*

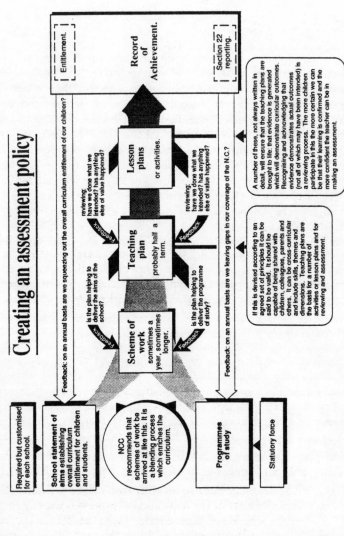

Creating an assessment policy

Required but customised for each school.

School statement of aims establishing overall curriculum entitlement for children and students.

Feedback: on an annual basis are we squeezing out the overall curriculum entitlement of our children?

Is the plan helping to deliver the aims of the school?

Scheme of work sometimes a year, sometimes longer.

Is the plan helping to deliver the programme of study?

Feedback

NCC recommends that schemes of work be arrived at like this. It is a blending process which enriches the curriculum.

Teaching plan probably half a term.

Feedback

reviewing: have we done what we intended? has anything else of value happened?

Lesson plans or activities.

reviewing: have we done what we intended? has anything else of value happened?

Entitlement.

Record of Achievement.

Section 22 reporting.

Programmes of study

Statutory force

Feedback: on an annual basis are we leaving gaps In our coverage of the N.C.?

If this is devised according to an agreed set of principles it can be said to be valid. It should be capable of being shared with children, colleagues, parents and others. It can be cross-curricular and include skills, themes and dimensions. Teaching plans are the basis for a number of activities or lesson plans and for reviewing and assessment.

A number of these, not always written in detail, will ensure that the teaching plans are brought to life. that evidence is generated which will demonstrate curricular outcomes. Recognising and acknowledging that evidence demonstrates actual outcomes (not all of which may have been intended) is a reviewing process. The more children participate in this the more certain we can be that their learning is confirmed and the more confident the teacher can be in making an assessment.

(This figure was devised by Cliff Jones and Joyce Humphreys (Liverpool Assessment Team) and Alan Wilkins (School Curriculum Industry Partnership).)

class/as a member of a group/as an individual/as homework or a combination of these);

- describe the intended learning outcomes (the ability to . . ., a knowledge of . . ., a willingness to . . ., an understanding of . . .);
- list the expected evidence which will show what has been learned (photographs/tapes/written/models);
- show how the work will be assessed (by the teacher/the student/ the teacher in consultation with the student/a group of teachers/a parent/ a work experience supervisor).

With an emphasis on explaining, sharing, reviewing and setting targets such a teaching plan connects with important elements of political education and citizenship. Acquiring the ability and confidence to make sense of their learning must help students to make sense of other aspects of life. It is argued below that effective curricular and assessment practice will make an important contribution to citizenship even if the subject content does not appear to be related to it. However, in the example of a format of such a teaching plan, given in Figure 8.2, the content is written specifically for citizenship.

In the *third stage*, teachers will devise lesson plans in accordance with their own styles of teaching and priorities. Clearly, however, the manner in which a teaching plan is transformed into the learning activities designed to bring it to life is as important as the written planning. It will make a crucial contribution to education for citizenship when the intentions of lessons, whether focused on specific learning outcomes or designed to recognise and acknowledge learning in whatever direction it takes, are shared and understood by both teachers and students so that collaborative review can take place.

The role of assessment in National Curriculum planning has been strongly emphasised for several reasons. Insofar as teachers thoroughly understand the relationship between curriculum planning and assessment *and* maintain control over this relationship, it is the contention of the present authors that this is the best safeguard of good educational practice within the constraints of the National Curriculum. But beyond this the argument would proceed that the development of student self-assessment is central to any notion of empowerment in education:

The issue here is a political one; that is, it is to do with the exercise of power. And power is simply to do with who makes decisions about whom . . . the objective of the [educational] process is the emergence of . . . a person who is self-determining – who can set his [sic] own learning objectives, devise a rational programme to attain them, set criteria of excellence by which to assess the work he produces, and assess his own work in the light of those criteria . . . assessment is the

Figure 8.2 *Teaching plan for study unit on political representation (part of a collection of units using cross-curricular themes to support NC Foundation subjects)*

LEARNING CONTEXT: as an individual, as a member of a small group and as a member of a whole class, the student will learn the main features of representative government in the United Kingdom. There will be opportunities to work with texts and to carry out a piece of research.

INTENDED OUTCOMES	ATTAINMENT TARGETS AND/OR LEVEL STATEMENTS	EXPECTED EVIDENCE
The student will demonstrate:		(a) labelled diagrams (1–9)
		(b) a wall chart (1–10)
a knowledge of	Eng 1–3	(c) questionnaire (6)
1. the qualifications for voting in the UK,		(d) results of survey (7, 8, 9b)
2. the different governmental bodies to which the electorate sends representatives,	Ma 9, 12, 13–15	(e) statement of learning
3. the frequency of elections,	Te 1, 3, 5	
4. the numbers voting (turnout) at different kinds of elections,	Sci 1	
5. the composition of Parliament in terms of, for example, race, gender and party;		*RESOURCES REQUIRED*
		text books, photocopier, possibly
the ability to		OHP, possibly time out of class.
6. devise a questionnaire suitable to discover frequency and consistency of voting,		
7. carry out a survey of at least ten voters, using the questionnaire,		
8. process and present the results;		
an understanding of		
9. how to find out about		
(a) different government bodies and		
(b) voting behaviour; and		
the experience of		
10. working independently and as a member of a group.		
11. reviewing the intended outcomes, together with the evidence, modifying and making more specific as appropriate in order to create a statement of learning.		

PROCEDURES FOR REVIEWING AND ACCREDITING

The intended outcomes of this learning unit are focused upon National Curriculum ATs listed above. After review by teacher and student in consultation a statement of learning will be made. This will record actual outcomes, some of which may be outside the National Curriculum. The statement may be placed in a ROA portfolio and contribute to the overall accreditation of the student.

most political of all the educational processes (Heron, 1981, pp. 55–63).

Self-assessment is a skill that must be developed like any other, and the starting-point is surely a plan for learning which can be shared and collaboratively reviewed by teachers and students. In the case of citizenship, discussion of the *criteria* for good citizenship is an essential element of the development of the good citizen. Whatever good citizenship is, it cannot be imposed; ultimately the citizen must set his or her own standards of behaviour and goals for action in the light of their understanding of the nature of citizenship and the clarification of their own values.

WHAT GOOD EDUCATION FOR CITIZENSHIP NEEDS

Returning to the theme of the prerequisites for good education for citizenship, a broad conception must encompass at least three elements – the qualities of the teachers, the ethos of the schools and the context of the wider society.

The teachers

Research carried out in the 1980s into political education in schools taught us that the problems of teaching this area of the curriculum are not those seen by the media and the Right of bias and indoctrination. Teachers were extremely sensitive to this possibility and were generally scrupulously fair in their presentation of material. Also, there is no lack of teachers committed to the importance of this area (see Brown 1990a, p. 41). The twin concerns of teachers were their lack of confidence in two key aspects – their *knowledge and understanding* of political issues and their *skills* in handling discussion of controversial issues which might stir up students' feelings, leaving them angry, upset and frustrated.

Undoubtedly a teacher who is personally interested in political and social issues and keeps him or herself well-informed will feel more confident to teach in this area. However, it is neither possible nor desirable for the teacher to be the sole (or even the main) source of information about *any* issue. Research skills are crucial for active citizenship and should be developed from the earliest age. These include critical reading, carrying out interviews and surveys, interpreting data, writing letters to request information, scientific analysis and experimentation. The teacher needs to bring two important qualities – awareness of the types and sources of data, how best to access them and what particular skills the students need to develop for this; and a sensitivity to the conceptual issues which will allow him or her to help students generalise from immediate local concerns to broader understanding.

This second requirement presents more of a problem than the requirement that the teacher should be well-informed. Conceptual understanding of social and political issues is most securely developed through training in one or more of the social science disciplines. The difficulties now faced by social science graduates wishing to enter the teaching profession, particularly at secondary level, have been noted and deplored elsewhere (Boxall and Burrage, 1989; Strivens, 1992). In the short term, there is a major role to be played by the subject associations (the Politics Association, the Economics Association and especially the Association for the Teaching of the Social Sciences) in supporting teachers who are not social science graduates to develop their understanding and confidence in these areas.

Lack of confidence about handling the discussion of controversial issues derives from a number of anxieties. First there is the issue of bias. What role should the teacher play? If he or she remains neutral, might the students be overinfluenced by strongly held views presented by their peers? How else can he or she ensure that balanced presentation of views required by the Education (2) Act of 1986 (Clauses 44 and 45)? Secondly, controversial issues generate emotions; can he or she handle students who become angry, upset or frustrated by the urge to action being blocked? Could he or she justify his or her teaching to parents and governors, if required? Confidence in this respect must ultimately derive from experience; but it is encouraging to note the requirement in the Council for the Accreditation of Teacher Education (CATE) regulations governing teacher training courses 'that students can teach controversial issues in a balanced way' (DES 1989, criterion 6.4).[7]

The school

Citizenship involves skills, and skills need to be practised in a relevant context. It also involves attitudes and values, and these are at least partially fostered through appropriate role models. The conclusion is unavoidable (and has been very widely recognised) that a basic requirement for good citizenship education is a democratic school ethos (Harber and Meighan, 1989; Strivens, 1986; Watts, 1980). In its simplest interpretation, this means that students experience their school as an institution where they have rights as well as duties, and a voice in determining the rules which constrain all members of this micro-society.

Socially this might be achieved through the setting up of a student council whose voice is taken seriously by all staff. At the least it has implications for the quality of the school's pastoral system. Academically the skills and values relevant to citizenship can be fostered by the adoption of flexible learning systems which give students increased autonomy and control over their own learning (Employment Department Group, 1990). At the least, the adoption of a model of curriculum planning and assessment as outlined above will help to create a supportive climate.

Society

Ultimately teachers have to remember that, even though schools may have a part to play in changing society, they also reflect society as it is. If citizenship is languishing in the wider society it will hardly flourish within the school. Bevan (1991), reflecting on the consultative conference called by the Speaker's Commission in February 1990 in Northampton, has some pertinent comments on this:

> It is not only the attitudes and conduct of the civil service and judiciary which raises doubts about a citizenship culture. It is . . . the ruthless extension of political patronage by people who claim they are ending a dependency culture and the mania for secrecy in government and media manipulation that makes one say that some people should apply the Northampton standards of citizenship to themselves (p. 3).

At the time of writing, the current Government has produced in short succession a Citizens' Charter, a Parents' Charter and a Patients' Charter, and has promised a Women's Charter. It is legitimate to ask if all these documents are designed to empower the people to engage with the real decision-makers, or rather to act collectively like a customer complaints department which separates government from the consumers. Since the charters have not been generated on a cross-party basis it will not be surprising if they are interpreted as Conservative propaganda rather than a genuine contribution to the development of a written constitution and bill of rights.

LOOKING TO THE FUTURE

Contemplating the current state of schooling in the United Kingdom, one has to say that the outlook for the establishment of good citizenship education is bleak. Pressure on the timetable squeezes the non-statutory aspects of the curriculum. Financial constraints on heads and governors will push them to give priority to staff appointments covering the statutory areas to the detriment of those trained in the social sciences. The need to look good in the league tables will raise the status of examined subjects in comparison with learning outcomes and experiences which rely upon the endorsement of the Record of Achievement.[8] More serious than these even, the control over assessment is being levered away from teachers, disempowering them as a professional group and making them less able to empower students through the crucial process of developing self-evaluation and self-assessment skills.

On the positive side can be counted some of the initiatives of the Training, Education and Enterprise Directorate (TEED) of the Department of Employment. These include the funding of flexible learning units; and Enterprise in Higher Education which is struggling to convince institutions of higher education that they should be fostering a broad range of skills for adult life beyond the narrowly academic. Sharing many of TEED's aspirations are the Royal Society of Art's (RSA) programme *Education for Capability* and *The Learning Society*.

All these initiatives seem a world away from most of the National Curriculum and schooling 5–16. Lister (1989) commented:

> part of the progressive potential of the National Curriculum lies in the fact that it may be viewed (and reformed) as *an entitlement curriculum* for all citizens.

Can this potential be activated? There are still many people inside and outside the education system who believe passionately in the importance of education for citizenship and will commit time and energy in pursuit of its realisation in the school system. As the National Curriculum takes shape in our schools, the developing role and status of the cross-curricular themes, skills and dimensions ought to be a key issue for education. More than any other area they are the elements which have the potential to transform our industrial performance; to strengthen our standing in relation to our European partners; to achieve a sense of identity and harmony within our multicultural society. They need all the protection and advocacy they can get.

NOTES

1. See also the bibliography on Citizenship and the Curriculum in *Social Science Teacher*, **20**, 3, Summer 1991, p. 103–4, and two editions of *Grassroots*, 56 and 57, the newsletter of the Politics Association.
2. The Manpower Services Commission has been through several manifestations and has been renamed in turn, the Training Commission, the Training

Agency and currently the Training, Education and Enterprise Directorate (TEED) of the Department of Employment. Although its function has varied, key personnel have remained the same.

3. HMI also showed an awareness of the extensive work which had been done in the area of political education, most significantly by the programme for political literacy at the Universities of York and London in the early 1970s. There is no reference to this background anywhere in the National Curriculum documentation or the Speaker's Commission Report.

4. It is not only National Curriculum testing which is becoming cruder with the substitution of 'pencil and paper' tests for the original SATs. Arguably of greater significance is the attack on GCSE coursework. Not only does this imply a lack of trust in teachers, it also narrows the range of learning outcomes which can be assessed and places great difficulties in the way of the development of 'vocational' GCSEs (see *Times Educational Supplement*, 18 October 1991). See also note 6 below.

5. The Department of Trade and Industry has handed over responsibility for some of its programmes under the Enterprise Initiative to TEED, for example 'Enterprise Awareness in Teacher Education'.

6. The relationship between Teacher Assessment (TA) and Standard Assessment Tasks (SATs) is similar to that between GCSE coursework and end-of-course examinations. The credibility of TA is ensured by a moderation process in which teachers share their intentions with colleagues and students and collaboratively review and assess learning. Constant reference to and testing of the criteria for assessment, as well as an appreciation of the context in which learning has taken place, enables teachers to make confident claims on behalf of students. SATs, as externally designed 'tests', are presented as a means of *confirming* the TA. If, however, they produce a result different from that of TA it becomes clear that they are in fact adjusters of TA; this despite the fact that they are different learning experiences, undertaken at different times and in different contexts. The central problem here appears to be that of trusting teachers. No matter how rigorous the TA moderation process, Government is only prepared to believe the results of what are increasingly likely to be short, simple and simultaneous tests. How is it possible to motivate teachers to invest time and trouble in good teaching plans leading to high quality teacher assessment if the results of those assessments in terms of levels are determined, maybe three or four years later, by a simple paper and pencil test?

7. It must be said however that the emphasis in CATE is on *balance* rather than *controversy*. In fact bias is more likely to arise from a policy of implicit *permeation* of citizenship throughout the curriculum than from planned, explicit discussion of issues by teachers properly trained to be aware of ideological positions.

8. The contrast in attitude of the DES and the Employment Department to the issue of 'quality assurance' of Records of Achievement is interesting. DES Circular 8/90 *A Record of Achievement* lays down the minimum legal requirement for reporting and looks at ways of enhancing that minimum by means of a RoA. In another document (DES, 1990) quality assurance in

RoAs is tied to the moderation of the procedures of National Curriculum Assessment. An Employment Department official speaking to the North West TVEI Nework on 14 October 1991 offered five 'keystones for quality assurance' of RoAs in schools:

- active involvement of students;
- a proper system of mentoring students;
- a code of practice;
- management support for RoAs; and
- external recognition of achievement by an accrediting body.

REFERENCES

Bevan, G (1991) 'Oh dear! Not Citizenship again: some reflections on the Northampton Conference on "Citizenship in Schools"', Grassroots (Newsletter of the Politics Association), 55 2–5.

Boxall, W and Burrage, H (1989) 'Recent, relevant experience: how CATE legitimates narrowly defined concepts of teacher education', *Journal of Further and Higher Education*, 13, 3, 30–45.

Brown, C (1990a) 'Personal and social education: a timetable ghetto or whole-school practice?' in B Dufour (ed) *The New Social Curriculum: A Guide to Cross Curricular Issues*, Cambridge: Cambridge University Press.

Brown, C (1990b) 'Active Citizenship – the wheel again?' (obtainable from University of York Department of Education documentation service).

Brown, C (1991) 'Education for Citizenship – old wine in new bottles?', *Citizenship*, 1, 2, 6–9.

Burrage, H (1990) Editorial, *Social Science Teacher*, 20, 2, 1.

Confederation of British Industry (1989) *Towards a Skills Revolution*, Report of the Vocational Education and Training Task Force, Sussex: CBI.

DES (1987) *Task Group on Assessment and Testing: a Report*, London: HMSO.

DES (1989) Circular 24/89 *Initial teacher training: approval of courses*, London: HMSO.

DES (1990a) Circular 8/90 *Records of Achievement*, London: HMSO.

DES (1990b) *Specifications for bidding under GEST 1990–91*, London: HMSO.

Employment Department Group (1990) *Flexible Learning: A Framework for Education and Training in the Skills Decade*, Sheffield: Employment Department.

Foster, S and Kelly, R (1991) 'Citizenship: perspectives and contradictions', *Talking Politics*, 3, 2, 73–6.

Harber, C and Meighan, R (1989) *The Democratic School: Educational Management and the Practice of Democracy*, Ticknall: Education Now Books.

Heater, D (1990) *Citizenship: The Civic Ideal in World History, Politics and Education*, London: Longman.

Heater, D (1991) 'What is citizenship?', *Citizenship*, 1, 2, 3–5.

Her Majesty's Inspectorate (1978) *Curriculum 11–16*, London: HMSO.

Her Majesty's Inspectorate (1983) *Curriculum 11–16: Towards a Statement of Entitlement*, London: HMSO.

Heron, J (1981) 'Assessment revisited', in Boud, D (ed.) *Developing Student Autonomy in Learning*, London: Kogan Page.

Lister, I (1989) 'Research on new initiatives in the field of social studies and citizenship education in England' (obtainable from University of York Department of Education documentation service).

National Curriculum Council (1989) *Curriculum Guidance 3. The Whole Curriculum*, York: NCC.

National Curriculum Council (1990) *Curriculum Guidance 8. Education for Citizenship*, York: NCC.

Speaker's Commission on Citizenship (1990) *Encouraging Citizenship*, London: HMSO.

Strivens, J (1986) 'Values and the social organisation of schooling' in P Tomlinson and M Quinton (eds) *Values across the Curriculum*, Lewes: Falmer.

Strivens, J (1992) 'The morally educated person in a multi-cultural society', in Modgil, S, Modgil, C and Lynch, J (eds) *Cultural Diversity and the Schools*, Volume 1, Lewes: Falmer Press.

Times Educational Supplement (1991) 'Vocational proposals threatened', 18 October 1991, p 1.

Trades Union Congress (1989) *Skills 2000*, London: TUC.

Training Agency (1989) *Training in Britain: A Study of Funding, Activity and Attitudes*, Sheffield: TA.

Watts, J (1980) *Towards an Open School*, London: Longman.

Chapter 9

Equal Opportunities within the National Curriculum

Paul Davies, Ann Mabbott and David Thomas

EQUAL OPPORTUNITIES WITHIN THE NATIONAL CURRICULUM

The aim of this chapter is to explore the complex and sometimes ambiguous nature of equal opportunities as presented within the National Curriculum framework. The structure is one which, following a general introduction, addresses race, gender and special educational needs, core and foundation subjects, religious education, alternative curriculum packaging and assessment. This is followed by a brief case study of one school's approach to equal opportunities.

The National Curriculum states a concern to promote the spiritual, moral, cultural, mental and physical development of pupils at school and of society itself. The National Curriculum is a preparation, it claims, for the 'opportunities, responsibilities and experiences of adult life' (DES, 1988a). The National Curriculum presents itself as an entitlement of every individual pupil to have a broad and balanced curriculum irrespective of sex, place or ethnic background. Such a framework would seem to imply a commitment to equal opportunities. Further support for this notion appears to come from other National Curriculum publications.

In October 1989, the National Curriculum Council issued *Circular 6* (NCC, 1989a), which deals with the context of the Education Reform Act (ERA) and cross-circular provision which it identifies as themes, dimensions and skills. The dimensions, the subject of this chapter, appear to be 'equal opportunities and education for life in a multi-cultural society'. Further, 'they require the development of positive attitudes in all staff and pupils towards cultural diversity, gender equality and people with disabilities'. Personal and Social Development (PSD) is placed within this section on dimensions in a rather ambiguous way. PSD, it says, should be promoted intentionally via the whole curriculum by means of the cross-curricular dimensions. Is PSD therefore a dimension in itself, or the outcome of the operation of the 'named' dimensions throughout the whole curriculum, including the core and foundation subjects?

Circular 6, sub-titled *Preliminary Guidance*, was followed by *Curriculum Guidance 3: The Whole Curriculum (CG3)* (NCC, 1990a). This maintains, in part 1, the links between the dimensions of equal opportunity and life in a multi-cultural society and PSD. Although *CG3* appears to develop and amplify *Circular 6* it throws up other issues. First, there is an emphasis on the dimensions in relation to access to the curriculum: 'Equal opportunities is about helping all children to fulfil their potential ... This means, firstly, ensuring that all pupils have access to the curriculum'. It describes equal opportunities as strategies for the delivery of the curriculum and details 'sex, social, cultural or linguistic background as factors influencing educational outcomes'. However, it makes no attempt to define the dimensions clearly nor to give these cross-curricular issues similar status to the foundations of curriculum planning, unlike the approach in Northern Ireland. Nevertheless it speaks of the need for a 'whole school' approach both to the dimensions themselves and to curriculum planning. This idea of a whole school approach is also evident in *CG2*, *'A Curriculum for All'* (NCC, 1989b).

NCC also produced *Information Pack Number 2* (1990) which clarifies some issues relating to the dimensions but still makes no attempt to define them further. However, it could be deemed to be highly supportive of action to secure equality of opportunity. The National Curriculum will help teachers to:

- assess what each pupil knows, understands and can do;
- use their assessments and the programmes of study to identify the learning needs of individual pupils;
- plan programmes of work which take account of their pupils' attainments and allow them to work at different levels; and
- ensure that all pupils achieve their maximum potential.

Circular 11 (NCC, 1991) also refers to issues of access, plus assessment needs and language in the classroom in relation to bilingual pupils. There is no clear statement of exactly what the dimensions consist of, no clear articulation of the relationship between the dimensions and the core and foundation subjects and their ambiguous relationship with PSD and their part in PSE. No further guidance is forthcoming. Indeed the report from the consultative committee on multi-cultural education has not and will not appear. Despite the apparent positive rhetoric of the documents, anxieties continue to be felt about the reality of equal opportunities.

To fulfil the requirements of ERA there is the need to address equal opportunities as a fundamental basis of planning, yet equal opportunities as a major dimension of the National Curriculum is ill-defined and of unclear status. It is left to teachers and schools to find their own way forward.

A value position for educators is to move society, on issues of race, gender, disadvantage and disability, away from discrimination, prejudice and inequalities. The National Curriculum makes it clear that the dimensions, whatever they are, are the responsibility of all teachers. It fails, however, to help them understand that responsibility or give them sufficient consistent guidance on how that responsibility needs to be enacted in their work.

All teachers are managers, of the curriculum and of pupils, and interact with both. They also interact with the community. However, all the significant documentation available to them focuses on none of these functions very clearly. Indeed, as will be illustrated, potentially it removes these functions from them. In our view, however, opportunities *do* exist for schools and teachers to successfully address equal opportunities issues, despite National Curriculum demands and ensuing constraints.

Throughout the whole process schools and teachers are engaged in making choices. For example, the National Curriculum could be interpreted as encompassing at least three different ways of packaging the curriculum. First, there is the 'grammar school' tradition of separate subjects, as indicated by the Government's focus on core and foundation subjects and on a hierarchy even within those. The existence in statute of a common curriculum for all pupils could be deemed as serving equality of opportunity; for example, those pupils who have not previously studied science will now do so. This, of course, begs the question, 'what kind of science?' While there possibly are constraints presented by the character of the programmes of study (PoS) and the attainment targets (ATs), choices remain for teachers in terms of the 'packaging' of these and the teaching and learning strategies used.

A second model could be the 'assessment objective-led model'. In this, opportunities are presented in terms of pupils achieving their potential and in terms of continuity and progression. However, the narrowness and inflexibility may constrain pupil achievement. If the teacher chooses to view statutory assessment as part of a pupil's achievements and emphasises the formative element – that is, individual differences can be taken into account – then equal opportunities could be addressed.

The third way might be described as the 'liberal' tradition. This allows teachers the opportunity to package the curriculum for their own particular pupils and it is claimed that this still exists under the National Curriculum. However, the predominance of the core-foundation subjects together with the assessment requirements may well constrain such an approach, especially the way in which they have been presented and their timing. *CG3* may give some occasion for hope:

> In due course it is likely that schools will 'throw all the attainment targets in a heap on the floor and reassemble them in a way which provides for them the very basis of the whole curriculum' (NCC, 1990a).

How realistic is this given the pressure upon schools and teachers? How far will they feel confident to make such changes in the light of the documentation of the National Curriculum? Each of these three interpretations offers, then, potential support for equal opportunities. Much will depend upon which predominates and the consequent choices and constraints.

CONTEXT – RACE, GENDER AND SPECIAL EDUCATIONAL NEEDS

It is possible to approach the matter of the context of race, gender and special needs in several ways. Each area could be treated separately in respect of some societal or nationwide perspective; it would be possible, for instance, to ask questions about the state of race relations in Britain in the 1990s. Similarly, questions could be asked about the framework within which the law provides for the conduct of people on racially sensitive matters, and the relationship between the legal position and the actual experiences of ethnic minority members of society's responses to them and the effectiveness of the law's redress and protection. Inquiries can be made to see to what extent schools are supportive of progressive intentions towards democratic/equalisation values.

Just as it is possible to provide an analysis at these three levels on race, so a similiar account can be provided on gender and special needs issues. The authors have debated on the advantages and disadvantages of prescribing a distinct and separate exegesis on race, gender or special needs as opposed to seeing these three 'areas' as exemplary, a set of shared concerns which can usefully be placed under a general cover of 'equal opportunities'. There are real benefits from having three separate kinds of discussion, each exclusively devoted to presenting its particular concerns and thus giving it a prominence and a place. The drawback of linking these three areas is that in philosophical terms in the politics of practicalities a broad grouping of 'equal opportunities' could mean that specific and legitimate concerns of interest groups are smothered in an all-embracing and vacuous rhetoric. Having faced these difficulties it was decided that at the level of schools, colleges and universities it was not possible to see how effective institutional policy development and improvements in practices, relationships and opportunities in one area could proceed without it having implications for the others.

In two of the areas – race and special needs – educators can use as significant reference points two major reports: Swann (DES, 1985) and Warnock (DES, 1978). The former, with its identification of parental concerns for the educational achievements of children from ethnic minority backgrounds drew attention to the need for whole-school curriculum policies for developing a curriculum appropriate to a pluralistic multi-cultural society within a framework 'of commonly accepted values, practices and procedures'. The ILEA (1985) report on special needs, suitably entitled *Educational Opportunities for All?*, notes that: 'an active policy of equal opportunities (is) essential in meeting special educational needs' (para. 2, 13 and 14).

Unlike race and special needs, gender issues cannot refer back to a major national report to provide a framework for policy and practices. Where is the gender equivalent of the Bullock Report (DES, 1975), that is, 'Gender across the curriculum?' Sensitivity to gender issues has emerged through multiple sources and a variety of agencies and inputs. Through books, HMI reports, research journals and active discussion, a consensus of sorts has emerged in which the education of equal opportunities for girls is not disputed as a curriculum or policy agenda item.

Legislation of a general nature exists relating to issues of race and gender, for example the Race Relations Act (1976) and the Sex Discrimination Act (1975); each of these has implications for the delivery of education. Specific education legislation, post-1941, gives little regard to issues of race and gender. There has been no national thrust with regard to either of these but there have been localised reponses in terms of LEA policy development, LEA, institutional and teacher activities. This has been well documented in terms of the variety and character of these responses as they have developed.

Additionally various HMI and DES publications have drawn attention to the need for race/gender issues to be considered in curriculum planning and school organisation. Many authors have written about the effects of race and gender on educational performance, access to the curriculum, behaviour in schools, etc.

Despite this there has been no national focus given to these issues.

While the Warnock Report (DES, 1978) paved the way for the abolition of the categories of handicapped children and development of a broader concept of special educational needs, neither it nor its legislative expression, the 1981 Act, were 'equal opportunity' texts. The ILEA (1985) Report, chaired by John Fish, was explicitly placing ILEA policy for special needs pupils within an 'equal opportunity framework':

> The broad thrust of these recommendations is to see special education needs (as defined by the 1981 Act) as one of a range of issues which schools are obliged to address (not as race, gender and social class issues) and not as a discrete or autonomous area divorced from the main life of the school and its curriculum concerns (Thomas, 1986, p. 101).

Given the extent and intensity of the education debate on matters relating to race, gender and special needs prior to 1987, it might have been a reasonable expectation that the National Curriculum proposals would have had some regard for them. Although the National Curriculum makes reference to them, there will seemingly be no specific support or documentation for these important dimensions. Why has this opportunity been missed?

In reflecting upon the marginalisation of equal opportunities in the legislation framework it seems there are several possible explanations:

- Equal opportunities was seen as a progressive and radical 'band-wagon' upon which a number of 'undesirable' educators and politics activists had boarded. That is, 'equal opportunities' was more 'left' than 'right'.
- It was seen as a 'distraction' from the main thrust of ERA and the National Curriculum which was to raise educational standards so that the nation could become more commercially competitive. That is, equal opportunities was a 'soft' issue; producing more and better technicians was the 'hard' issue.
- Equal opportunities in these terms has been and will continue to be a highly controversial issue on which there are many views as well as many pressure groups not always sharing the same agenda.
- The personalisation and politicisation of the race/gender issues have not allowed equal opportunities to be a central professional issue in education.
- The lack of any clear definition, beyond exhortation to provide a broad, balanced relevant 'curriculum for all' irrespective of race, gender, place or special need, inevitably meant that the dimensions of the National Curriculum would be left in a grey area and remain problematic.

- The very idea of equal opportunities was never emotionally grasped by the intellectually 'dead, white male' power brokers of the ERA.

Subsequent to the Consultation Document (DES, 1987) attempts have been made by NCC and the Schools Examination and Assessment Council (SEAC) to ensure that equal opportunities issues are present within the whole curriculum. Equal opportunities has been marginalised by its additive status to the National Curriculum's basic scaffold. The compulsory common curriculum with its side-lining of race, gender and special needs may actually be a barrier to social reform and could leave the semi- and informal structure of schools unchanged. Standardised assessment 'tests' (remember when they were called 'tasks'?) have the potential for exacerbating gender and ethnic divisions, creating what Arnot (1989) called 'the national hidden curriculum'. The use of the 'whole curriculum' and cross-curricular themes, dimensions and skills represents a belated and inadequate response to these issues.

It is well understood that within the ERA there is an unresolved paradox. Section 1 of the Act spells out in broad terms the notion of all-round development for pupils, while Section 2 focuses upon the foundation subjects which can only deliver a portion of the broad aims. However, the prominence, status and significance of the foundation subjects is such as to make clear the differential value which the Act attaches to the narrow and the broad curriculum.

Attempts to resolve the paradox have been directed towards making clear that the National Curriculum does not provide the whole of the broad and balanced regime required and by attempts to reinforce cross-curricular activities. However, in spite of these efforts the real politics of the Act is that the National Curriculum is predominantly a subject-specific curriculum. The attendant programmes of study, attainment targets and assessment procedures are there with the force of law while the cross-curricular aspects are required but not statutory – a distinction which is not missed by schools hard pressed to accommodate the National Curriculum let alone the whole curriculum. Where the cross-curricular themes are part of the National Curriculum programmes of study they are legally binding, but where they are not built-in they have the status of a desirable but unenforceable recommendation.

How are teachers and schools to respond? Professional experience indicates a wide spectrum of responses. At one end those schools which already had an awareness of and commitment to a well defined positive standpoint on equal opportunities have been able to critically appraise the various messages in National Curriculum literature on race, gender and special needs and incorporate National Curriculum recommendations into their own curriculum. Some schools have used the tensions and paradoxes relating to equal opportunities in the National Curriculum to raise the issues for the first time. Some have assumed that the very existence of a common core curriculum ensures that equal opportunities are being delivered. Yet others have responded by producing a school rhetoric to match the National Curriculum rhetoric but no more. For some institutions the low priority and status given to equal opportunities, together with inconsistencies in the National Curriculum have reinforced their own passivity.

Some schools have grasped the opportunity offered by National Curriculum planning to provide strategies for equal opportunities; others have been diverted or even overwhelmed by the apparent constraints. All have made or are making choices about it. Here lies the paradox: the dimensions of the National Curriculum appear to be non-statutory and yet to deliver the statutory requirement of the broad-based relevant curriculum for all requires equal opportunities to be a (the?) major issue in planning.

The intention is to argue that it is within the choices exercised by teachers and schools that equal opportunities provision will either be enhanced or denied; and ultimately a curriculum delivered which will give all pupils access to the knowledge, skills and competencies to which they are entitled.

The authors have made a tactical decision to consider race, gender and special needs as areas which can be conjointly considered under an equal opportunities umbrella. There is something specifically distinctive about special education needs.

The Consultation Document [DES, 1987] which presented the broad intention of the ERA showed that special educational needs were not at that time high on the Government's list of educational priorities (the Special Educational Needs Task Group was set up in February 1989). In spite of attempts to get special needs issues addressed Section 1 (2) which gives the legislation's intention to provide a broad and balanced curriculum to promote the rounded development of pupils did not refer to 'all' pupils. And while the word 'all' is in Circular 5/89 (DES, 1989a), circulars do not have the force of law. *From Policy to Practice* (DES, 1989b) makes clear that the National Curriculum is not the whole curriculum, but those parts of the Act dealing with ways in which the National Curriculum requirements may be modified or disapplied shows that these only apply to the National Curriculum and not to the whole curriculum. As Wedell (1989) noted, this implies that only the National Curriculum is safeguarded within the law.

Sections 17–19 of the ERA dealing with modifications and disapplication of the National Curriculum for pupils with special educational needs are presented as part of the 'flexible' character of the National Curriculum. Among the obvious areas of modification are those to meet the learning needs of pupils with sensory or physical handicaps, although there are problems in translating such good intentions into practice. Disapplication could be regarded as more contentious as far as equal opportunities is concerned.

Assessment within the National Curriculum either by SATs or Teacher Assessment (TA) presents significant challenges to educators where pupils with special educational needs are concerned. Finding assessment practices in terms of modes of presentation, operation and response which allows such pupils to show what they have achieved is no easy matter. How pupils with fragile emotional and behavioural states will cope with testing is a matter of some concern.

Part of the rationale for assessment is the provision of public information which will enable the community to judge the quality of its schools. It has been decided that the performance of pupils with special educational needs do not have to be included in a school's published results. In other words, how a school responds to the needs of its pupils with learning difficulties is discounted! It implies that schools need not take too seriously the commitment to deliver a balanced curriculum to all pupils.

The fragile position of equal opportunities for those with special educational needs is further undermined by the Local Management of Schools (LMS) and the opting-out facility. A current debating point for special educators is whether it is LMS or the opting-out clause which will have the greatest negative impact on the developments which were in hand in the post-Warnock, pre-ERA period. Between-school competition, the decline in the range and quality of support services, the relative value placed by school financial managers on expenditure on special needs projects compared with other school activities, all seem to offer a significant threat.

While some of the more insidious implications of the National Curriculum have been deflected by the attention given to special educational needs by the subject working groups, it should be noted that the documents from the working groups show a difference between the early reports of (core subjects) and the later ones, which tend to give more attention to special needs and other equal opportunities issues.

It is a conviction held strongly by many that the 1988 Act runs counter to the spirit of the 1981 Act and will have the effect of reversing much of the gains made during that period. Among the pessimistic forecasts are an increase in the numbers of pupils brought forward for statementing (the process by which a child's special educational needs and the appropriate resources to meet those needs are identified) and an increase in the numbers of pupils, especially at secondary phase, who will be disaffected from their schooling, together with an increased demand for special school places with a concomitant growth of the private special school sector.

Much of the debate about the impact of ERA and the National Curriculum on special educational needs was conducted on the basis of speculation, or informed guess work, but with the latest HMI report (1991) some evidence is now available. From visits to 26 ordinary and 50 special schools, HMI found evidence of much good practice in a period when teachers have been under severe pressure. This is a token of the qualities of our teaching force. Among the issues addressed by HMI is that of the improvement being made by special and mainstream schools to enrich and extend the curriculum experiences of their pupils. Examples are cited of special schools extending experiences through the introduction of a modern language. However, there is little in the report about the relationship between the National Curriculum, the whole curriculum and the values teachers hold about what is desirable for special pupils.

> In ordinary schools, integration of pupils with statements into subject classes usually gave them access to a broader curriculum, a wider range of resources and the stimulus of working with their peers. Some primary and special schools were having difficulty in deciding whether or not the National Curriculum reasonable time requirements were being fulfilled because elements of both core and other foundation subjects were being subsumed within project or topic work. Some special schools had difficulty in making justifiable links between some existing activities, for example independence training and other life skills, and the emerging National Curriculum orders. Those schools that had clear aims and objectives about the whole curriculum, including the National Curriculum, were more capable of making such links. (HMI, 1991, p. 16).

HMI does not address directly cross-curricular issues but it is clear that pupils with special educational needs do not appear to have been negatively affected by the National Curriculum provision; indeed there are many encouraging and positive aspects to the report. It is tempting to regard this as having happened because of the commitment of teachers rather than an intended and planned outcome of the ERA.

However, the fact that 'special needs' is placed within the legal framework of the ERA suggests that for its drafters 'special needs' had a greater valence than did 'race' or 'gender'. The paradox of an entitlement curriculum for 'all' is an intriguing one in that there are opportunities for exemption and modification.

Returning to the opening statement on values and equal opportunities it follows that if equal opportunities is taken seriously, what the legislation is stating is less important than schools' and individual teachers' attitudes. By 'teachers' we mean all those in education regardless of status, who need to live these values. From this perspective, equal opportunities, whether in curricular, assessment or human relationship terms becomes an issue for each person and for their institution as a whole, not forgetting that the institution is embedded in a community context.

In essence this places the discourse on curricula, assessment and relationships into a moral and ethical dimension. The enterprise of education requires teachers, at all levels, to reflect on the moral and ethical principles which they hope guide their practice. For us a frame would be based around our beliefs in such key concepts as justice and equality and would not be content with policies or practices which seek merely to identify liberal, pluralistic, value-free areas of concern and make them available for student debate within structures which were proclaiming the reverse. Rather it may be asserted that equal opportunities is an area for political action. It is an inescapable logic that equal opportunities requires social action (individual/institutional). The socially just school will go beyond multi-culturalism to anti-racism, beyond concern for gender to anti-sexism, beyond special needs to individual needs.

CORE AND FOUNDATION SUBJECTS AND RE

Are the National Curriculum's concerns about equal opportunities solely about access to the SATs, that is, access to a particular discipline and at particular levels? Should not the entitlement curriculum mean that the core/foundation subjects and ensuing attainment levels are some, from among a variety of means by which the pupil becomes the adult social being? This latter view would seem to be supported by the liberal tradition (packaging the curriculum in the light of a school's decisions about its values and hence its aims and objectives). In the light of the emphasis upon core/foundation subjects, their assessment and National Curriculum training, this approach seems very distinct. Indeed the appearance of guidance on the whole curriculum (*CG3*) only after several subject 'ring binders' had appeared, indicates a lack of coherence. In addition it must be noted that the foundation (and core) is still shifting, with recent amendments to mathematics and science documents further inhibiting whole-curriculum planning, suggesting the focus still remains on the subjects. The statutory nature of these subjects and their assessment increases the pressure on schools and individual teachers when making choices about the nature and purpose of education within their own context.

The preliminary, or interim, subject documents from the subject working groups exhibit vast differences in their attention to and interpretation of equal opportunities issues and their stances, although the need to address matters of race, gender and special educational needs was in all the remits. The English document, for example, (NCC, 1989c) devotes separate chapters to bilingual children (Ch. 10), equal opportunities (Ch. 11) and special needs (Ch. 12). Technology, (NCC, 1989d) has a heading 'Equal Opportunities', divided into three: 'Pupils with Special Needs', 'Gender' and 'Ethnic Minorities'. Mathematics (NCC, 1988) has four separate sections: 'Individual Pupil Needs', 'Equal Opportunities', 'Ethnic and Cultural Diversity' and 'Special Educational Needs'.

The terms themselves are therefore used consistently in three major documents. Other interim documents can be deemed to be strong on equal opportunities issues. Modern Foreign Languages (NCC, 1990c), for example, has a chapter (Ch. 4) ranging over a number of related issues stemming from 'It must not be assumed, however, that simply exposing pupils for five years of learning will in itself ensure equal opportunity to benefit from the experience' (14.2), and goes on to consider two vital factors: how pupils are taught and what they are taught. These are examined in detail in relation to equal opportunities.

The final orders, the statutory parts of the curriculum, consist of the ATs, statements of attainments and the programme of study. They are accompanied by non-statutory guidance. It is difficult to find traces of the working parties' thinking on equal opportunities issues within the statutory parts of the documents. Within the programme of study of a number of subjects, equal opportunities threads can be traced to a greater or lesser extent. In History (DES, 1991a) for example, in the general requirements for Key Stages 2–4, 'Pupils should be taught about the social, cultural, religious and ethnic diversity of the societies studied and the experiences of men and women in these societies'. This, of course, is extremely problematic given the nature and content of some of the programme of study. One might look for similar ones in other subjects. The setting out of each is different, making tracking difficult. In English (DES, 1989c) in relation to the programme of study supporting AT2 (reading), there are specific mentions, for example, of 'stories from other cultures' and, supporting AT2, 'Listening and responding to stories, rhymes, poems and songs . . . include examples from different cultures . . .'. Pupils should be encouraged to respect their own language(s) or dialect(s) and those of others. In science (DES, 1989d) in the programme of study material supporting the original 409 statements of attainment there are few discernable threads: 'Children should consider similarities and differences between themselves and other children' in relation to the programme of study.

Where guidance occurs it is in the non-statutory part of the binder, which always appears after and separately from the statutory part. Once again there are variations in approach. Technology (DES, 1990) quotes from its own interim document its views in relation to equal opportunities issues thus preserving its thinking for those coming new to the document. Under the heading 'Meeting the needs of all pupils', B1 non-statutory guidance (NSG) it deals separately with SEN, providing equal opportunities and cultural diversity. Science (DES, 1989e) deals with the issues under a heading 'Teaching Science' (NSG A8–A10) but without separate headings. English (DES, 1989c) has, under 'Planning Schemes of Work' (C1) paragraphs on SEN and bilingual children in general planning considerations and some further sentences in 'Aspects of Programme of Study' (D1–2). Mathematics NSG (DES, 1989f) has no discernible equal opportunities section beyond 'the mathematical development of each pupil is different . . .', under the heading 'The order of activities should be flexible' (B8). The only section here referring to any cross-curricularity is on F1 'a rationale for cross-curricular work', which is about the contribution of mathematics to the whole curriculum. History (DES, 1991b) and Geography (DES, 1991c) both have specific sections in their NSG. History, in C18–20 puts together equal opportunities and multi-cultural education, examining them under programme of study, AT and resources. It examines separately 'Catering for Individual Needs' but not under the same headings as equal opportunities. Multi-cultural geography heads up 'Cross-curricular elements – dimensions' (C18).

Information technology is a National Curriculum cross-curricular skill as well as being an AT within the Technology Orders. No guidance has yet been issued related to cross-curricular skills. However, in the Technology NSG there is ample evidence of serious consideration being given to special needs and it uses the term 'equal opportunities' to embrace both race and gender. It progresses beyond the pious to giving realistic and helpful advice. In relation to special needs, it is noticeable that this advice suggests information technology could be the only route by which some pupils can access their entitlement. In a convenient and handy form this document also gives teachers a summary of the main research finding on the educational correlates of race and gender.

Within the core and foundation subjects, therefore, there is a diversity of usage of terms like equal opportunities, diverse approaches and a lack of consistent guidance even in the appearance of the appropriate sections, etc. Primary/special school teachers using all the documents will have great difficulty in responding consistently across the curriculum to the dimensions. Secondary teachers in the main using one only will find little help either.

In relation to the cross-curricular elements CG3 specifies Science AT17 which 'may cast light on ways in which scientific ideas are affected by the social, moral and spiritual context in which they develop' (DES, 1989d, p. 2). In the latest version of the Science proposals AT17 has disappeared without trace!

The mathematics interim document (DES, 1987) makes it quite clear that mathematics is to be regarded as a neutral discipline in the old tradition in its sections on individual need, equal opportunities, ethnic and cultural diversity and SEN (10.10–10.32). In referring to materials/ examples free from gender bias and giving 'as much attention to the girls in their class as to the boys', they say that this 'is a matter of good teaching practice, rather than something to be taken on board in the construction of the mathematics curriculum itself' (10.17). Or, in relation to ethnic and cultural diversity it is sometimes suggested that the multi-cultural complexion of society demands a 'multi-cultural approach to mathematics ... [this] could confuse young children'. Therefore no 'multi-cultural' aspects appear in any of the attainment targets. Most telling in 10.10/10.11: 'Is there a risk that the existence of prescribed national targets will make life more difficult for the low achieving child by reinforcing his or her sense of failure?' This need not be so. It is perhaps an act of faith but we believe that the existence of clear attainment targets for mathematics ' . . . can help many children who are currently low achievers . . .'. This hardly represents a pedagog-ically sound approach to the curriculum!

The interim report from the physical education (PE) working group (NCC, 1990d) contains some extremely positive statements: *'teachers should treat all children as individuals with their own abilities, difficulties and attitudes'* (2.36) (emphasis in the original).

> The diversity of the activities which it is possible to learn in physical education should allow all children, regardless of their sex, educa-tional need, cultural or ethnic background, access to the skills, knowledge and understanding which form the foundation for the physically educated person (NCC, 1990e, 2.37).

Crucially in 2.38: 'it is important to emphasise that *mere access cannot be equated with real opportunity'* (emphasis in the original). It then lists ten key issues particular to PE in relation to equal opportunities and says they 'will be given fuller treatment in the final report' (2.39). It is hoped that this is indeed so. However, given the previous examples it will quite likely be located in the non-statutory guidance. The recently published proposals for PE (DES, 1991d) make 'equal opportunities' a vital part of its curriculum philosophy. It asserts that 'equal opportunities' must be given the status of a 'guiding and leading principle' (p. 15).

The prominence given to equal opportunities in the PE proposals is matched by that given in Music (DES, 1991e) and Art (DES, 1991f). The proposals from the art working group articulate the belief that 'Art is for All' (Ch. 11) and deals with this concept through the following divisions: 'Ethnic minorities', 'Special educational needs', 'Gifted children', 'Gender' and 'Aspects particular to Wales'. The proposals from the Music Working Group examine equal opportunities through an overlapping but not identical set of categories: 'Cultural heritage and diversity', 'Special educational needs' and the 'Musically very able'. These latest documents suggest that the concept of equal opportunities is widening the definition and interpretation of the two dimensions.[1]

This movement is reflective of the discourse which has already taken place in Northern Ireland where 'Education for mutual understanding' and 'Cultural heritage' have the status of educational 'themes' (NICC, 1990). In the Northern Ireland documents it appears that the 'themes' have a status not far below that of the 'six areas of study' and it is 'intended that the content of each educational theme should be incorporated in the Programme of Study of appropriate subjects and the objectives met through the medium of those subjects' (p. 3). This could be interpreted as meaning that the cross-curricular themes will be delivered through the areas of study (subjects) which will probably have a major say in curriculum and assessment choices. It is also significant that the prominence given to 'mutual understanding' and 'cultural heritage' should emerge in Northern Ireland where such concepts are not abstractions but a felt need with life and death connotations.

One of the clearest statements of the power and potential of cross-curricular planning may be found in the Curriculum Council for Wales' *A Framework for the Whole Curriculum 5–16 and Developing the Whole Curriculum* (CCW, 1991). It advises that the National Curriculum and religious education do not form the whole curriculum and pupils have needs which these subjects alone will not satisfy. It recommends that in planning a broad and balanced curriculum 'schools should consider the curriculum as comprising eight aspects of learning to which pupils are entitled'. These 'aspects', which include 'Expressive and aesthetic', 'Linguistic and literary' and on through to 'Technological' are seen as appropriate sites for curriculum planning and development 'considered as a whole'. These aspects 'are inherently cross-curricular' and subjects will contribute to more than one aspect. To these aspects CCW seeks to integrate 'Themes, Competencies and Dimensions'. Each aspect is brought forward as a planning centre where the aspects' principle features are defined and the major contributions of the statutory subjects, the contributions of 'other subjects and activities' and the contribution of themes, competencies and dimensions are given. To have 'Equal opportunities, Cultural diversity and Special needs' placed close to the centre of curriculum planning, where there is virtually no assumption that any one subject will dominate an aspect, seems to be a potentially fruitful planning model. It is reminiscent of the lines in CG3 where it was suggested the ATs might be 'thrown in a heap on the floor and reassembled'. However, here, there is an attempt to provide a coherent philosophy for the 'reassembly'.

Religious Education

The provision under ERA is that religious education (RE) stands alongside the core and foundation subjects as part of the entitlement of all pupils. This compulsory element is to be 'in the main Christian while taking account of the teaching and practices of the other religions represented in GB' (DES, 1988a, 5–8). A daily act of worship is required for all; it should reflect 'the broad traditions of Christian belief without being distinctive of any particular Christian denomination' (6.5). Local Standing Advisory Councils on Religious Education (SACRE) are required and have the power to remit the 'broadly Christian' requirement of worship in certain instances. Although RE is compulsory it is not nationally but locally defined and assessed, based upon a locally agreed syllabus approved by SACRE. Thus it is in danger of becoming marginalised and being a token religious education rather than truly spiritual enhancement. This is particularly so since a number of prominent faith groups cannot be well represented in the composition of SACRE or even represented at all. Secular groups, of course, have no representation. Already, in a number of places, action has been taken by Christian groups or individuals to challenge schools' wide interpretation of 'broadly Christian'. The potential of the spiritual in the education process may be further diminished by lack of cooperation from groups excluded from the discussions.

The challenge is to develop RE in ways which are sensitive to believers but also develop in non-believers empathy and understanding.

ALTERNATIVE PACKAGING

Core and foundation subjects represent the tradition of packaging and curriculum via various subject disciplines. Each of these has its own statement of attainment and ten levels within which each vary considerably in character across the subjects.

For instance, within the Geography Orders (DES, 1991g) the statements of attainment range from what might be considered clear objectives, for example 'State where they live' (AT2 1C) and 'use 4 figure coordinates to locate features on a map', to statements of attainment which would seem subjective or vague, for example, 'synthesise patterns, relationships and processes in the home region' (AT2 10a) and 'review environmental problems arising from the development of industry in the USA, USSR or Japan' (AT1 7a).

Opportunities do exist for teachers to repackage the programmes of study. This is encouraged by National Curriculum documentation, inviting ownership of the subject matter. The nature of the programmes of study, however, presents constraints which, in some subjects, are particularly straitjacketing. Within the Geography Orders the burden of a heavy content and the placing of location study within the key stages makes teacher ownership potentially problematic. However, choice still exists. Similarly, the emphasis on core units with a potentially Anglo–European bias within the History Orders (DES, 1991h) presents a major constraint. This has been exacerbated by pronouncements emphasising a content for history which focuses on concepts which many pupils may find difficult to relate to, for example, national political movements. The opportunity to relate historical events to pupil reality and understanding has not been encouraged by the dismissal of events of the last 20 years as 'current affairs', not history. However, the choice still exists for teachers to reclaim the delivery of history units by focusing on what relates best to their pupils; the ways to or entry points into the subject matter may not be recognised by either the statutory programme of study or corresponding statements of attainment.

Opportunities do exist to interpret the programme of study in ways which support equal opportunities aims and issues. Although the History Orders focus on a potentially white Anglo–centric historical experience, teachers can, through the choices they make, counter this imbalance. For instance, in Key Stage 2, pupils studying Britain since 1930 can consider 'religious changes and their effect on everyday life' and 'cultural changes' within the context of multi-faith, multi-cultural approach. Where the National Curriculum encourages the study of non-European society (for example, in geography where pupils study a location in an economically developing country or in history where a non-European society is to be studied) teachers have the choice of reinforcing stereotypes and prejudices or countering them. Attention to the quality of resources will be critical here. Selection and interpretation are key issues for the teacher. If pupils' only experience of black history is 'Black people of the Americas: 16th Century to early 20th Century' then the teacher must make a judgement about the suitability of that unit. If it is chosen, will it focus on black achievement or will black people be condemned to the status of historical victims?

ASSESSMENT

Assessment in education needs to consider why we assess, what should be assessed, how to do it, the interpretation of obtained results, responding to and feeding back and communicating results with whose who have a reason to access them. Assessment has multiple purposes: selection, keeping up standards, motivating learners, supplying feedback (to students and teachers) and providing formative and summative data.

As already stated, the major aim of equal opportunities is to move away from discrimination, prejudice and inequalities. Assessement, according to NCC Training Pack 2 (NCC, 1990b) would seemingly support this:

- assess what each pupil knows, understands and can do;
- use their [teacher] assessment and programme of study to identify the learning needs of individual pupils;
- plan programmes of work which take account of pupils' attainments and allow them to work at different levels;
- ensure that all pupils achieve their maximum potential.

This is reflective of the apparent support for equal opportunities principles in both the TGAT report (DES, 1988b) and the two subsequent supplementary reports, published in March 1988. For example: 'We recommend that the basis of the national assessment system be essentially formative' (DES, 1988b, para. 17) and the national system should employ tests for which a wide range of modes of presentation, operation and response should be used so that each may be valid in relation to the AT assessed' (para. 50). And yet the TGAT report devotes only two paragraphs (51.52) to the issue of gender and ethnic bias in assessment and the submission of the Equal Opportunities Commission is an Appendix (F) the details of which are discussed neither in the main report nor in the supplementary reports. There are no detailed discussions about assessment in relation to ethnic issues.

It is claimed that as most pupils with special educational needs will take regular standard attainment tasks the SATs designers will not include in their test/task anything which will put special educational needs pupils at a disadvantage, 'if it can possibly be avoided' [CG2 p. 12]. The possibility of extra time for the completion of a SAT by pupils with particular difficulties has been considered. The chief recommendations to ensure that pupils with special educational needs are not disadvantaged during assessment are the adaptations of modes of presentation (the ways tasks are presented), modes of operation (the ways the task can be carried out) and the modes of response (how the pupils may present their work). SEAC (1989) notes the desirability of structuring work into small steps and that teacher record-keeping systems need to be adapted to record progress towards attainment targets. Most crucial of all for pupils with special educational needs is the need for continuous assessment, not just at key stages. It is difficult to over estimate the crucial role of sensitive assessment policies and practices for special educational needs pupils.

The rhetoric is about the process and the diagnosis of individual progress/needs. The reality seems to be: summative testing not only at 16 as TGAT envisaged, but also at 7, 11 and 14; SATs which have become progressively narrower in character and the current pressure for 'simpler pencil/paper-type tests' over a shorter period of time. All this moves away from assessment as a part of normal classroom practice and impinges upon the teachers' choice of tasks in relation to their knowledge about their pupils. Can teachers plan assessment which accommodates the range of individual needs or will they prepare the whole class for particular testing? If the latter, there may be a danger that the range of assessment tasks be narrowed to become confirmatory of predictions of performance. This could also have a knock-on effect on the programme of study and choices about syllabus content, as already indicated. The result could be an assessment-led curriculum which is not diagnostic but a 'labelling-predictive' one, fuelled by notices about 'league tables', teacher appraisal and parental choice. Good practices could be abandoned to accommodate this approach.

The considered view of the authors is that teachers can still exercise choice and reclaim the process for meeting individual needs. Our experiences show that where there has been open dialogue and mutual support within schools, with moderators and with local education authorities, the assessment process at Key Stage 1 has been supportive of equal opportunities principles. This process has involved:

- planning to start the programme of study with assessment as an integral part, the programme of study being focused on pupil needs, not on assessment needs;
- the programme of study planning, at its earliest stage giving

recognition to issues of race, gender, SEN and social class;
- the active positive participation of pupils/parents in the whole process;
- the identification of opportunities to assess pupil achievement other than by SATS, both in the core and foundation subjects and in other areas of the curriculum;
- the recording and reporting of such achievements in as positive a way as possible by the Record of Achievement.

Thus individual achievements have been recognised and built upon, beyond the narrower statutory requirements. This has been achieved by the professionalism of the teachers concerned in claiming ownership of the tasks. It has of course been achieved by much hard work, effort (and personal expense at times!) and the potential denial of equal opportunities by assessment has been avoided.

A WHOLE SCHOOL APPROACH TO EQUAL OPPORTUNITIES

CASE STUDY

School X is a mixed 11–16 comprehensive school. It operates on a split site and its catchment area is almost entirely made up of council house estates.

The school has identified the examination of equal opportunities as one of the issues it would tackle in the school development plans (SDP) for 1988–90. In this case, initially 'equal opportunities' certainly equated with gender issues. As with other items identified in the SDP, a small working group of interested staff was convened to plan the equal opportunities responses. Some members of the group were there by virtue of their role in the school, for example, the staff development/in-service coordinator or the TVEI coordinator, others by virtue of a specific interest (a newly qualified teacher of sociology or the head of the fifth year) and the head also served on this group. On those occasions when one of the authors was present, by invitation, the group was six strong, five being male all of whom had significant responsibilitiese and status within the school.

From the outset they decided that any consideration of equal opportunities should be approached via an equality model, that is to say that all staff should have the opportunity to contribute to any decisions made. To achieve this, they argued, would require:

1. time to be allocated to any activities, given the plethora of other demands upon teachers;

2. recognition to be given to the fact that there would be differing degrees of awareness about gender issues amongst the 55 staff;
3. that the issues should be raised in relation to the school's stated aims and objectives, and thus as part of every teacher's professional concern.

It was therefore agreed that an in-service day be given over to equal opportunities (gender) issues, which would address points 2 and 3 above, and that particular parts of 'directed time' (as it was then) be allocated for departmental participation.

The group felt it was appropriate to hand over the organisation and content of the day to a person external to the school. Having failed to secure the services of the equal opportunities Commission they invited the LEA teacher fellow for equal opportunities to plan and provide the day within the parameters above.

The day itself was divided into four parts:

- An input from the leader on general issues related to gender and the curriculum; this was focused using statements from the school's own literature and related to research findings, statistics etc. A second input came from the head of a neighbouring school about their own findings on gender-related issues.
- Mixed groups of staff were given the same real statistical information related to option choices, examination results, etc.; the groups were asked to produce a series of responses to the statistics in the light of the first session. These were brought back and shared with the whole group. A composite list of issues was drawn up by the two leaders acting as facilitators.
- The same groups as before took this list and prioritised it in relation to their own school. Each group had to produce a reasoned case for their first three choices agreed upon by all group members. All the groups then returned to share this activity and as a result a list of six priorities was identified.
- The groups each took one priority and tried to produce an action plan which would facilitate action over a short time scale.

It should be noted that the head of the school concerned had underlined in his introduction to the day that the activities and plans were not to be regarded as hypothetical but would indeed form part of the school's development over the next few months. Second, there was a particular focus throughout the day that permitted the discussion/activity to be related to individuals, departments, faculties and pastoral groups as well as the school as a whole. Among the outcomes were:

- raised awareness for some staff on gender issues;

- participation by all staff with different colleagues;
- school priorities arrived at by discussion/agreement; and
- a commitment to action.

A further outcome was a growing realisation of the relationship of gender issues with other issues about access to the curriculum, for example, SEN and race. These arose in the course of a number of discussions throughout the day.

Since this in-service day occurred, one year ago, the school has produced a pamphlet detailing research on gender issues by 15 departments. It is noticeable that many had involved pupils in their research, most had related it to their general curricular planning and all had identified further action points. In addition, a more general model for whole-school planning was developed:

- School's own contexts as well as statutory requirements and other initiatives – lead to choices.
- Values and attitudes – leads to choices.
- Aims and objectives – leads to choices.
- Curriculum planning, including planning for National Curriculum commitment and action – leads to choices.
- Current situation.
- Desired situation – leads to choices.
- Necessary changes – leads to choices. .
- Priorities (corporate/departmental/individual); time scales; action plans (corporate/departmental/individual) – lead to choices.

At each stage equal opportunities issues must be addressed, for example, who participates in the process? What organisational arrangements are made? How are resources allocated? And so on.

CONCLUSION

It has been argued that while equal opportunities issues and concerns are formally present in National Curriculum documentation and legislation, their stature, significance and definition is ambiguous. However, this position is not wholly without possibilities. These include a belief in the ways in which individual teachers, schools and communities, working within the ERA/National Curriculum framework can find ways of regaining and retaining ownership of this area of the curriculum.

Within the over-arching form of the National Curriculum and national assessment, there are many areas which could be taken over by teachers in the interests of equal opportunities. There are choices both at the level of curriculum strategies and at the level of radical choices. Pedagogic styles and human relationships infused with a commitment to equal opportunities offer ways in which individual teachers can make their choices matter. Equally schools, through their equal opportunities policies, can create an intellectual and moral climate in which the presence of equal opportunities is integrated with the fabric of curriculum decision-making, assessment procedures and whole-school planning.

The down side is that such progressive approaches depend more on individual teachers, subject departments, and heads exercising their initiative than being centrally supported by legislation or official NCC/SEAC documentation.

It is clearly impossible to guess the outcomes. Experiences suggest that viewed nationally, a diverse and patchy picture may well emerge with the emphasis and prominence given to equal opportunities being ultimately in the hands of schools rather than in a national commitment centrally recognised and resourced.

NOTE

1. However, the NCC Consultation Reports on Mathematics and on Science (both September, 1991) draw attention to the perceived inadequate treatment of special needs, gender and ethnic and cultural diversity within these core subjects. For example, the mathematics report states that 32 per cent of respondents felt that there was a missed opportunity to pay attention to the issue of pupils with special education needs, ethnic and cultural diversity (46 per cent) and equal opportunities (42 per cent). In science the comparable figures were 'over a third' who thought the issues were 'not properly addressed' (special needs) and there were similar serious concerns over ethnic and cultural diversity and equal opportunities. Specific mention was made of the loss of AT17 which was seen as not helping the cause of equal opportunities.

REFERENCES

Arnot, M (1989) 'Crisis or challenge? Equal opportunities and the National Curriculum', *NUT Education Review*, 3, 2, 7–13.

CCW (1991) *A Framework for the Whole Curriculum 5–16 and Developing the Whole Curriculum*, Cardiff: CCW.

DES (1975) *A Language for Life*, The Bullock Report, London: HMSO.

DES (1978) *Special Educational Needs. Report of the committee of enquiry into the education of handicapped children and young people* (The Warnock Report) Cmnd 7212, London: HMSO.

DES (1985) *Education for All. Report of the committee of inquiry into the education of children from ethnic minority groups* (The Swann Report) Cmnd 9453, London: HMSO.

DES (1987) *Mathematics Working Group. Interim Report*, London: HMSO.

DES (1988a) *The Education Reform Act 1988*, London: HMSO.

DES (1988b) (TGAT) *National Curriculum Task Group on Assessment and Testing – A report*, London: HMSO.

DES (1989a) *The Education Reform Act 1988: The School Curriculum and Assessment*, Circular No. 5/89, London: HMSO.

DES (1989b) *National Curriculum; From Policy to Practice*, London: HMSO.

DES (1989c) *English in the National Curriculum*, London: HMSO.

DES (1989d) *Science in the National Curriculum*, London: HMSO.

DES (1989e) *Science in the National Curriculum (Non-statutory Guidance)*, London: HMSO.

DES (1989f) *Mathematics in the National Curriculum (Non-statutory Guidance)*, London: HMSO.

DES (1990) *Technology in the National Curriculum*, London: HMSO.

DES (1991a) *History in the National Curriculum (England)*, London: HMSO.

DES (1991b) *History in the National Curriculum (Non-statutory Guidance)*, London: HMSO.

DES (1991c) *Geography in the National Curriculum (Non-statutory Guidance)*, London: HMSO.

DES (1991d) *Proposals. Physical Education*, London: HMSO.

DES (1991e) *Music for Ages 5 to 14*, London: HMSO.

DES (1991f) *Art for Ages 5 to 14*, London: HMSO.

DES (1991g) *Geography in the National Curriculum*, London: HMSO.

DES (1991h) *History in the National Curriculum*, London: HMSO.

DES and the Welsh Office (1987) *The National Curriculum: A Consultation Document*, London: HMSO.

HMI (1991) *National Curriculum and Special Needs; Preparations to Implement the National Curriculum for Pupils with Statements in Special and Ordinary Schools, 1989–90*, HMI Report, London: HMSO.

ILEA (1985) *Educational Opportunities for All?* Report of the committee reviewing provision to meet special educational needs (The Fish Report), London: ILEA.

NCC (1988) *Proposals. Mathematics*, York: NCC.

NCC (1989a) *Circular Number 6: The National Curriculum and Whole Curriculum Planning: Preliminary Guidance*, York: NCC.

NCC (1989b) *Curriculum Guidance 2. A Curriculum for All – Special Educational Needs in the National Curriculum*, York: NCC.

NCC (1989c) *Proposals. English for Ages 5 to 16*, York: NCC.

NCC (1989d) *Interim Report. Technology in the National Curriculum*, York: NCC.

NCC (1990a) *Curriculum Guidance 3. The Whole Curriculum*, York: NCC.

NCC (1990b) *The National Curriculum Information Pack No. 2*, York: NCC.

NCC (1990c) *Proposals. Modern Foreign Languages*, York: NCC.

NCC (1990d) *Physical Education. Interim Report*, York: NCC.

NCC (1991) *Circular Number 11: Linguistic Diversity and the National Curriculum*, York: NCC.

NICC (1990) *The Northern Ireland Curriculum. A guide for teachers*, London: NICC.

SEAC (1989) *Recorder No 2*, London: SEAC.

Thomas, D (1986) 'Special Educational Needs; translating policies into practice', *Journal of Curriculum Studies*, 18, 1, 100–101.

Wedell, K (1989) 'Children with special educational needs and the National Curriculum', *NUT Review*, 3, 2, 32–6.

Index

Index

Index